Encyclopedia of the Animal World

MAMMALS
The Hunters

Christopher O'Toole & John Stidworthy

Facts On File
New York • Oxford

Distributed by
World Book, Inc.

THE HUNTERS
The Encyclopedia of the Animal World:
Mammals

Managing Editor: Lionel Bender
Art Editor: Ben White
Designer: Malcolm Smythe
Text Editor: Miles Litvinoff
Assistant Editor: Madeleine Samuel
Project Editor: Graham Bateman
Production: Clive Sparling

Media conversion and typesetting:
Peter MacDonald and Una Macnamara

AN EQUINOX BOOK

Planned and produced by:
Equinox (Oxford) Limited,
Musterlin House, Jordan Hill Road,
Oxford OX2 8DP

Prepared by Lionheart Books

Library of Congress
Cataloging-in-Publication Data
O'Toole, Christopher.
Mammals: the hunters/Christopher O'Toole
and John Stidworthy.
p.96, cm.22.5×27.5 (Encyclopedia of the
animal world)
Bibliography: p.1
Includes index.
Summary: Introduces predatory members
of the mammal family, from leopards and
cheetahs to whales and dolphins.

1. Predatory animals – Juvenile
literature. 2. Mammals – Juvenile
literature [1. Predatory animals. 2.
Mammals.] I. Stidworthy, John 1943-.
II. Title. III. Series.

QL758.076 1988   599.053-dc19
88-16933

ISBN 0-8160-1959-2

Published in North America by
Facts on File, Inc.,
460 Park Avenue South,
New York, N.Y. 10016

Origination by Alpha Reprographics Ltd,
Harefield, Middx, England

Printed in Hong Kong

10 9 8 7 6 5 4 3 2

FACT PANEL: Key to symbols denoting general features of animals

SYMBOLS WITH NO WORDS

Activity time

● Nocturnal

● Daytime

◐ Dawn/Dusk

○ All the time

Group size

◨ Solitary

▦ Pairs

▧ Small groups (up to 10)

■ Herds/Flocks

◪ Variable

Conservation status

☠ All species threatened

☠ Some species threatened

No species threatened (no symbol)

SYMBOLS NEXT TO HEADINGS

Habitat

■ General

◣ Mountain/Moorland

◢ Desert

≈ Sea

■ Amphibious

◺ Tundra

◸ Forest/Woodland

● Grassland

≋ Freshwater

Diet

■ Other animals

■ Plants

◪ Animals and Plants

Breeding

◎ Seasonal (at fixed times)

◡ Non-seasonal (at any time)

CONTENTS

PREFACE

The National Wildlife Federation

For the wildlife of the world, 1936 was a very big year. That's when the National Wildlife Federation formed to help conserve the millions of species of animals and plants that call Earth their home. In trying to do such an important job, the Federation has grown to be the largest conservation group of its kind.

Today, plants and animals face more dangers than ever before. As the human population grows and takes over more and more land, the wild places of the world disappear. As people produce more and more chemicals and cars and other products to make life better for themselves, the environment often becomes worse for wildlife.

But there is some good news. Many animals are better off today than when the National Wildlife Federation began. Alligators, wild turkeys, deer, wood ducks, and others are thriving – thanks to the hard work of everyone who cares about wildlife.

The Federation's number one job has always been education. We teach kids the wonders of nature through *Your Big Backyard* and *Ranger Rick* magazines and our annual National Wildlife Week celebration. We teach grown-ups the importance of a clean environment through *National Wildlife* and *International Wildlife* magazines. And we help teachers teach about wildlife with our environmental education activity series called *Naturescope*.

The National Wildlife Federation is nearly five million people, all working as one. We all know that by helping wildlife, we are also helping ourselves. Together we have helped pass laws that have cleaned up our air and water, protected endangered species, and left grand old forests standing tall.

You can help too. Every time you plant a bush that becomes a home to a butterfly, every time you help clean a lake or river of trash, every time you walk instead of asking for a ride in a car – you are part of the wildlife team.

You are also doing your part by learning all you can about the wildlife of the world. That's why the National Wildlife Federation is happy to help bring you this Encyclopedia. We hope you enjoy it.

Jay D. Hair, President
National Wildlife Federation

INTRODUCTION

The *Encyclopedia of the Animal World* surveys the main groups and species of animals alive today. Written by a team of specialists, it includes the most current information and the newest ideas on animal behavior and survival. The Encyclopedia looks at how the shape and form of an animal reflect its life-style – the ways in which a creature's size, color, feeding methods and defenses have all evolved in relationship to a particular diet, climate and habitat. Discussed also are the ways in which human activities often disrupt natural ecosystems and threaten the survival of many species.

In this Encyclopedia the animals are grouped on the basis of their body structure and their evolution from common ancestors. Thus, there are single volumes or groups of volumes on mammals, birds, reptiles and amphibians, fish, insects and so on. Within these major categories, the animals are grouped according to their feeding habits or general life-styles. Because there is so much information on the animals in two of these major categories, there are four volumes devoted to mammals (*The Small Plant-Eaters; The Hunters; The Large Plant-Eaters; Primates, Insect-Eaters and Baleen Whales*) and three to birds (*Waterbirds; Aerial Hunters and Flightless Birds; Plant- and Seed-Eaters*).

This volume, *Mammals – The Hunters*, includes entries on lions and tigers, wolves, foxes, bears, weasels and civets, as well as seals, dolphins, porpoises and Sperm whales. Together they number some 380 species. These animals feed almost entirely on other animals. Many of them may seem to be fierce and frightening. But some, for example the dolphins, seem remarkably gentle and affectionate towards people, and even use a system of communication that parallels our speech.

Scientists call meat-eating animals carnivores. Most carnivorous mammals that live on land belong to a group called the Carnivora, which includes the cats, dogs, bears, raccoons, weasels, mongooses and hyenas. In the Encyclopedia all these animals and their relatives are dealt with in this volume. However, also included here are those members of the quite unrelated pouched mammals (marsupials) that share many of their hunting characteristics.

Carnivores that hunt in the sea fall into two groups. First, the seals, sea lions and walrus, which spend much of their time at sea catching fish and other animals. These are called the Pinnipedia. Second, the so-called Toothed whales (dolphins, porpoises, Sperm whales), which hunt fish and other large prey. Another group of whales, the Baleen whales, are dealt with in volume 4.

Each article in this Encyclopedia is devoted to an individual species or group of closely related species. The text starts with a short scene-setting story that highlights one or more of the animal's unique features. It then continues with details of the most interesting aspects of the animal's physical features and abilities, diet and feeding behavior, and general life-style. It also covers conservation and the animal's relationships with people.

A fact panel provides easy reference to the main features of distribution (natural, not introductions to other areas by humans), habitat, diet, size, color, pregnancy and birth, and lifespan. (An explanation of the color coded symbols is given on page 2 of the book.) The panel also includes a list of the common and scientific (Latin) names of species mentioned in the main text and photo captions. For species illustrated in major artwork panels but not described elsewhere, the names are given in the caption accompanying the artwork. In such illustrations, all animals are shown to scale; actual dimensions may be found in the text. To help the reader appreciate the size of the animals, in the upper right part of the page at the beginning of an article are scale drawings comparing the size of the species with that of a human being (or of a human foot).

Many species of animal are threatened with extinction as a result of human activities. In this Encyclopedia the following terms are used to show the status of a species as defined by the International Union for the Conservation of Nature and Natural Resources:

Endangered – in danger of extinction unless their habitat is no longer destroyed and they are not hunted by people.

Vulnerable – likely to become endangered in the near future.

Rare – exist in small numbers but neither endangered nor vulnerable at present.

A glossary provides definitions of technical terms used in the book. A common name and scientific (Latin) name index provide easy access to text and illustrations.

LION

It is dawn on the East African plains. A herd of wildebeest and zebras graze in the pale sunlight. Suddenly, the early morning quiet is shattered by the roar of a lion. But the zebras and wildebeest carry on feeding peacefully. They have nothing to fear. The lion is a male and his fearsome roars are not directed at them. Instead, they are a signal to other males: "This is my home area and these are my females. Keep away!" For the grazing animals, danger will come later in the day. Then, a group of lionesses begin their prowl in search of animals to kill and eat.

LION *Panthera leo*

● ■

● Habitat: grasslands of E. Africa, desert areas. Some in India.

■ Diet: antelope, gazelle, warthogs, wildebeest, zebra, smaller animals.

◖ Breeding: litters of 1-5 after pregnancy of 100-119 days.

Size: head-body 7½-11ft; weight 270-530lb; males larger.

Color: light tawny, belly and inside legs white, backs of ears black, mane of male tawny to reddish or black.

Lifespan: 15-24 years.

The lion's strength and haughty expression have led people to call it the "King of Beasts". Like all the cats, the lion has a sleek, muscular body with a deep chest. The short powerful jaws are well-armed with a fine set of sharp teeth, designed for chewing and tearing meat and even for cracking open bones. The feet have a set of powerful claws and, together with keen hearing and sight, a lion is superbly equipped for the hunting life.

Male lions are heavier than females (lionesses) – sometimes half as big again. Being larger enables the male to push his way between females at a kill and get at the best meat. Males sometimes steal carcasses killed by other animals, but mostly they feed on animals killed by the females.

Only the male lion has a mane of long hair on the head and shoulders. This makes him look larger and fiercer than he really is, and is useful in arguments with other males – a smaller male will retreat before starting a fight. If there is fighting, the thick mane protects its owner against the teeth and claws of a rival. The main role of male lions is to defend the home area of the family group and to protect females from other groups of male lions.

THE MIGHTY HUNTER
Male lions rarely hunt. Most hunting is carried out by lionesses. The prey consists of large animals such as gazelles, antelopes, warthogs, zebras and wildebeest. Lions also kill and eat lizards, birds and smaller mammals such as rats. An adult male lion needs about 15lb of meat a day, while the smaller female requires 11lb each day.

Lionesses usually hunt together. Several females stalk and spread out to surround a prey animal. They try to get as close as possible, using long grass or bushes as cover. A lioness can run as fast as 35mph, but some of its prey can run much faster than this, so a slow, quiet approach is just as important as speed.

Only one in four charges by lionesses ends with a kill. After being knocked to the ground, the animal is

▼This male lion has killed a horse which strayed from a farm. Now, using all his might, he drags it to cover.

◄Two males from the same pride groom each other. They are probably brothers or half-brothers, and grooming maintains the bond between them.

▼Family life in a resting pride of lions. A lioness (1) suckles three cubs, only one of which (2) is hers. The others belong to females (3) and (4). Two resting males (5) do not mind the playing cubs (6).

killed by a bite to its throat, which breaks the windpipe, or by having its jaws clamped shut by the lioness. Either way, it chokes to death.

Although lions are good hunters, up to three-quarters of the animals they eat are killed by hyenas and stolen by the lions. People usually think of hyenas as being scavengers, but in fact lions are more so.

A PRIDE OF LIONS

Lions are the most social of all cats. They live in groups called prides. A pride usually has 4 to 12 adult females and their young, with 1 to 6 adult males. The females are usually related to each other, most often as cousins. The males are also related to each other – mostly as half-brothers – but not to the females.

A pride has its own home area or territory. The size of this area depends on how many prey animals it contains. A pride of lions lets other prides know where its territory ends by patrolling the boundary, roaring, and marking it with urine at regular places. Although both sexes will defend territories against intruding males, it is the males of a pride that do most of this.

Lions first breed when they are between 36 and 46 months old, and females can breed several times a year. Within a pride, there is little fighting between males over females. Instead, the first male to meet a willing female mates with her.

At between 109 and 119 days after mating, a lioness gives birth to a litter of 1 to 10 cubs. The cubs stop suckling milk from their mother at about 6 months old, although they start eating meat earlier than this. As many as three-quarters of all lion cubs die of starvation before they reach 2 years of age.

LIONS AND PEOPLE

For thousands of years, people have respected lions for their strength and bravery. Many royal families in Europe had lions on their flags and coats-of-arms.

Because lions are thought to be so fierce and strong, people could show how brave they were by killing a lion. Many hunters went to Africa from Europe and North America to kill lions and bring back their heads or skins as trophies.

Nowadays, the hunting of lions in Africa is strictly controlled and most visitors go to photograph animals instead.

Although there are still many lions in Africa, they are under some threat. When scrub is cleared for farming,

◄A group of resting lionesses on a rocky outcrop. Lions spend much of their time sleeping, especially during the heat of the day and after a heavy meal. But these seven lionesses have woken suddenly after being disturbed and are very alert.

prey becomes scarce and lions may then disappear from the area, or may be shot if they begin to hunt farm animals.

Lions and their prey animals are both threatened when the vegetation in the areas they inhabit changes. Lions used to be found all over northern India, the Middle East and Africa north of the Sahara Desert. As the deserts increased in size and encroached on scrubland, the lions disappeared.

Up to 2,300 years ago, lions lived in Greece, and cave paintings show that in much earlier times the lion was widespread in most of Europe. The last lions in the Middle East were wiped out by hunting about 100 years ago. Today, lions live only in Africa south of the Sahara and in one forest nature reserve in north-west India.

MAN-EATERS

The Romans imported lions from North Africa and used them to kill prisoners as a kind of public entertainment. Many early Christians were killed in this way. Lions, though, are not normally man-eaters. Stories of man-eating lions in Africa usually result from old or sick lions attacking humans because they are easier to catch than normal prey.

Sometimes, though, healthy lions will eat people if their supply of game has been reduced by farming or other human activities. A famous example took place in the late 1800s, when the railway between Kenya and Uganda was being built. So many workmen were killed by a pair of lions that work on the railway had to be stopped until the lions were shot.

If lions are to survive, we must ensure that they and their game animals have plenty of space. This is provided by the great game parks and nature reserves of Africa, but even here lions are under threat as human numbers increase and the need for farmland becomes greater.

TIGER

A lone tiger pads softly through the dappled sunlight of an Indian forest. Silently, he picks his way through the undergrowth, stopping at frequent intervals to sniff the air. Perhaps he is searching for a female on heat or for a suitable animal to prey on. Soon he will take to a stream to cool down in the heat of the day. There, he drinks and may rest, or simply wait for an unwary deer to come for a drink.

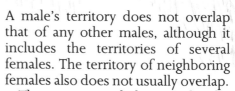

1

2

Tigers are found only in Asia, where they live in forests with plenty of cover. They are the largest living cats and, unlike lions, are solitary hunters. There are eight races or subspecies of the tiger, some of which are now extinct. All surviving races are endangered, despite the establishment of tiger reserves in India.

The tiger is well armed for its life as a stalk-and-ambush hunter. The hind legs are longer than the forelegs, for powerful leaping, and long sharp claws on the front feet enable it to grasp and keep hold of struggling prey. The tiger eats whatever it can catch, but most of its prey are medium to large-sized animals, including wild pigs and deer.

HOME RANGES

Each tiger has its own home range or territory. Those of females (tigresses) are about 8sq miles in area, while male territories are from 24 to 38sq miles.

A male's territory does not overlap that of any other males, although it includes the territories of several females. The territory of neighboring females also does not usually overlap.

The tiger regularly patrols the borders of its home range. It marks the borders with urine mixed with a scent from the anal gland, which it sprays on to trees, bushes and rocks. It also deposits droppings throughout its area.

For a female tiger, owning a territory has advantages. She gets to know the area well and discovers the best places to find prey. Having control over the prey in her area is important

▼ Camouflage – dark stripes on a pale background break up the body outline of a tigress lying in ambush.

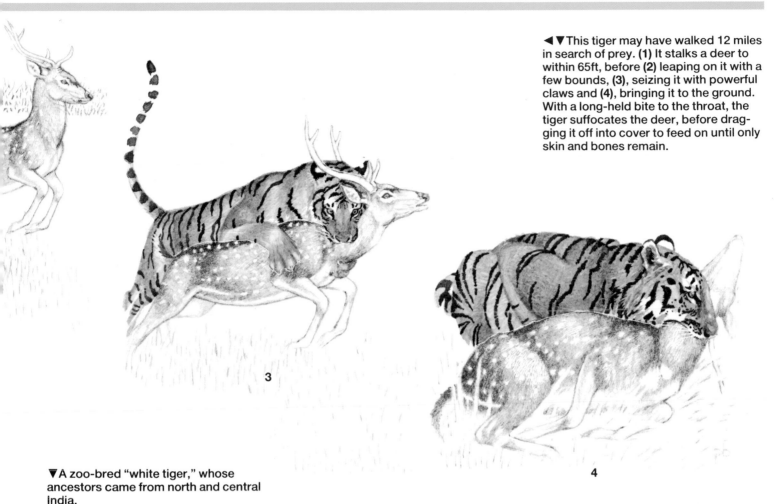

◄▼This tiger may have walked 12 miles in search of prey. **(1)** It stalks a deer to within 65ft, before **(2)** leaping on it with a few bounds, **(3)**, seizing it with powerful claws and **(4)**, bringing it to the ground. With a long-held bite to the throat, the tiger suffocates the deer, before dragging it off into cover to feed on until only skin and bones remain.

3

4

▼A zoo-bred "white tiger," whose ancestors came from north and central India.

if she has cubs to look after. For a male, with his much bigger territory, access to prey is probably not so important. His advantage lies in being able to monopolize the females living within his borders.

SOLITARY MOTHERS

Tigers begin to breed when they are 3 to 4 years old. They mate at any time of the year in the tropics, but only in winter further north. A female has a litter of three or four cubs, each weighing about 2lb. The cubs live in a den until they are 8 weeks old, after which they follow their mother around.

The female looks after the cubs until they are 18 months, and they may remain in her territory for 2½ years before leaving to find their own home ranges.

LEOPARDS AND JAGUAR

A loud, raucous rasping noise comes from near the top of a thorn tree in East Africa. It is a male leopard, signalling his dominance over the area he surveys from his prickly perch, high above the ground. At night he will leave this tree and prowl in search of prey.

There are three species of leopard, the largest and most familiar being the ordinary African leopard. This beautiful spotted cat lives in all of Africa south of the Sahara and extends through Asia as far east as China.

It is the most widespread member of the cat family and can live in almost any habitat so long as there is some cover. This solitary hunter is mainly active at night. Using a combination of silence, speed and cunning, it preys on a wide range of small mammals and birds. Leopards may ambush their prey or stalk it before making a final rapid pounce. Because of its wide choice of small prey, the leopard avoids competition with lions, tigers, African wild dogs and hyenas, all of which hunt larger animals.

Like other large cats, the leopard is territorial. Each animal has its own home range and it will not allow other leopards of the same sex to come close to it. The home range of a female leopard is 4 to 12sq miles and overlaps very little with the territories of neighboring females. Male territories are larger and include one or more female territories.

Both sexes defend their areas by fighting, and scent-mark widely by

LEOPARDS AND JAGUAR Felidae (4 species)

● ◩ ☠

Habitat: most areas with good cover, from tropical rain forest and dry savannah to cold mountains.

Diet: very varied, from fish and birds to small- and medium-sized mammals.

Breeding: litters of 1-6 after pregnancy of 85-110 days.

Size: head-body 2-6½ft; weight 33-250lb; males often much larger than females.

Color: black to black-ringed brown spots of varying size on a fawn to pale brown background.

Lifespan: up to 12 years, 20 years in captivity.

Species mentioned in text:
Clouded leopard (Neofelis nebulosa)
Jaguar (Panthera onca)
Leopard (P. pardus)
Snow leopard (P. uncia)

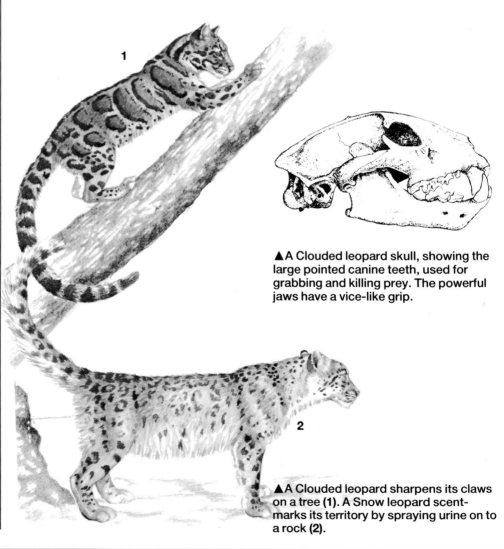

▲A Clouded leopard skull, showing the large pointed canine teeth, used for grabbing and killing prey. The powerful jaws have a vice-like grip.

▲A Clouded leopard sharpens its claws on a tree (1). A Snow leopard scent-marks its territory by spraying urine on to a rock (2).

▲ Making a larder. This leopard drags its prey to a tree, storing it out of reach of scavengers.

▼ Relaxed and unconcerned, a leopard dozes in an acacia tree, shaded from the midday Sun.

spraying urine on to tree trunks, branches and logs. They also signal their presence by calling with a saw-like rasping sound.

RAISING YOUNG
The leopard begins to breed when it is about 2½ years old. Females come into heat every 3 to 7 weeks, and mating occurs at all times of the year. The male plays no part in rearing his offspring.

The female produces up to six blind, furry cubs, each weighing between 1 and 1½lb. She keeps her young hidden in a safe place for the

first few weeks, and they soon begin to follow her around. They depend on her for 18 to 20 months, after which she mates again. Female cubs may then take over part of their mother's home range, while male cubs leave when they are 2 or 3 years old and set up separate territories of their own.

HUNTED FOR FUR
Although there are well over 100,000 African leopards still in the wild, the species is endangered. Farmers sometimes kill leopards because of their attacks on livestock and many are

hunted for their fur, which is highly prized for making coats and jackets for rich but unthinking women. In Africa the leopard is still hunted for "sport" by Western visitors. In parts of Asia leopards sometimes become man-eaters, and some individuals have been known to kill over 100 people.

Although the leopard is probably declining in numbers in many parts of its range, attempts to save it can be successful. In the deserts of Israel the number of leopards has actually increased.

RARE AND MYSTERIOUS

The Snow leopard or ounce lives in Asia. It is a rare and shy inhabitant of mountain country, living in steppe and coniferous forest at altitudes of 6,000 to 18,000ft. Here it preys on a wide range of animals, from birds and mice to marmots, musk deer, wild sheep and ibex, which it follows as they migrate higher up the mountains during summer. In winter the Snow leopard seeks lower ground, where it preys on hares, wild boar, gazelles and deer. Each Snow leopard has a large territory of up to 40sq miles. It stalks its prey and pounces from a distance of 15 to 50ft.

During the breeding season, from January to May, the Snow leopard gives up its solitary life, and males and females hunt together in pairs. Before giving birth, in spring or early summer, the female finds a safe den which she lines with her own fur. She produces up to four cubs. They are blind at birth, then open their eyes after several days. By about 2 months they are active and playful, remaining with their mother throughout their first winter.

At the approach of winter, the fur of the Snow leopard becomes thicker and this helps it survive the intense

▶The "black panther" is really just an all-black leopard and is not a separate species.

◀Larger than the Old World leopard, this jaguar surveys the scene from a look-out high up in a tree.

cold. In summer the fur becomes finer again. Sadly, its wonderful coat may well be the Snow leopard's downfall – this cat is now very rare because of the illegal trade in its fur.

Also rare is the Clouded leopard, which lives in dense forests in India and South-east Asia. Although heavily built, it is the smallest of the leopards. Birds, squirrels and monkeys form much of its prey. This beautiful cat spends most of its time in trees, where it is a skilled climber.

The Clouded leopard is rarely seen in the wild, and nothing is known about its social life.

GOOD SWIMMER

The jaguar is the largest cat found in the Americas and is similar to the leopard in its way of life. It lives in forests, swamps, and even deserts, from the south-western United States south to Patagonia.

The jaguar preys on fish, frogs, turtles and their eggs, birds, caymans, rats, mice and larger animals like capybara, deer and monkeys. It is a good swimmer and prefers to live near water.

Except during the mating season, the jaguar is solitary. Each has a territory of between 2 and 200sq miles. For some unknown reason, a jaguar may travel up to 500 miles each year. Jaguars breed from about 3 years of age. A female gives birth to two to four blind cubs, each weighing between 1½ and 2lb. After about 13 days, the eyes open; the young remain with their mother for about 2 years.

◀An alert Snow leopard patrols its remote mountain territory. It may climb to above 21,000ft after prey.

CHEETAH

A female cheetah sits on top of a termite mound in Africa, while her three cubs frolic below. For the moment, she is content to let them play. But she will soon carry on teaching the cubs how to hunt and kill prey. They will follow her white-tipped tail through the long grass, stopping at her command before she chases down a gazelle. Then she will call them to the prey. While she holds it down, they will practise the killing bite.

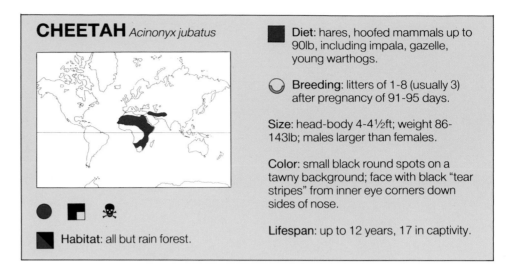

CHEETAH *Acinonyx jubatus*

Diet: hares, hoofed mammals up to 90lb, including impala, gazelle, young warthogs.

Breeding: litters of 1-8 (usually 3) after pregnancy of 91-95 days.

Size: head-body 4-4½ft; weight 86-143lb; males larger than females.

Color: small black round spots on a tawny background; face with black "tear stripes" from inner eye corners down sides of nose.

Lifespan: up to 12 years, 17 in captivity.

Habitat: all but rain forest.

▲Powered by long, muscular hind legs, a cheetah sprints at 56mph. A flexible spine allows it to make very long strides. The claws, even when held back, are not covered by a sheath but exposed, giving it a better grip on the ground.

▼With the acceleration of a high-powered sports car, a cheetah breaks cover to chase a Thomson's gazelle. An average chase covers 560ft and lasts less than a minute. About half of these chases end with a successful kill.

With its small head and slim, loose-limbed build, the cheetah is the most distinctive of the big, spotted cats. Once common over much of Africa and the Middle East, only about 25,000 now survive in Africa.

FASTEST ON EARTH
The cheetah is best known as the fastest animal on Earth, capable of speeds up to 56mph. This skill is put to good use when it pursues hares and its more normal range of prey species, which include impala and gazelles.

Every year in Africa, the female cheetah follows the migration of its

◄Alerted by the presence of prey, a female cheetah leaves cover and is about to begin stalking.

prey animals, moving through a home range of up to 320sq miles. The female is not territorial, so the home ranges of two or more females may overlap.

The male cheetah is aggressive. He will fight other males, sometimes to the death, in defense of his territory. Males usually live in groups, remaining together for life and marking their territories by spraying urine at regular intervals on landmarks such as tree stumps. Males also hunt together.

The cheetah breeds all the year round. The male does not help with rearing the young. At birth the cubs weigh 9 to 11 ounces, and their eyes open at 2 to 11 days. When a few

weeks old, they leave their hiding-place and follow the mother around, eating some of the prey she catches. She weans them at about 3 months, but they remain with her until they are 17 to 23 months old, when female offspring leave one by one. The young males leave as a group – they are usually chased away from the area by older, more experienced males.

SECRETS OF SUCCESS
The cheetah shares the same areas as lions, leopards and hyenas. Its legendary speed helps it survive in this fierce competition. The cheetah also tends to hunt around the middle of the day, when the other large animals are usually asleep.

Daytime hunting, though, has its disadvantages. Vultures may drive a cheetah away from its kill and, at the same time, attract the attention of other hunters, which may then steal the prey. The cheetah avoids this by taking the kill to a hiding-place.

▼A strangling bite to the throat has killed this Thomson's gazelle. The cheetah now drags it to cover.

WILD CATS

In a remote forest in the far north of Scotland, a solitary European wild cat steals through the night. Despite the darkness, he knows his area well, guided by landmarks scented with his own urine. Always on the prowl by night, seeking food or a mate, he is alert to danger and whatever opportunities he comes across. The faint rustling of a mouse attracts his attention. Quietly and patiently he stalks it. Then, with a rapid pounce, he has it in his jaws, and the mouse is dead.

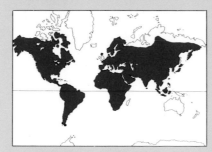

WILD CATS Felidae
(*28 species*)

Size: smallest (Leopard cat): head-body 14in, weight 7lb; largest (puma): head-body 6½ft, weight 225lb.

Color: varying patterns of dark spots or stripes on pale gray, fawn or tawny background, coat sometimes unpatterned; black "tear stripe" often running from eye.

Lifespan: up to 12-15 years (European wild cat).

Species mentioned in text:
African wild cat (*Felis sylvestris lybica*)
Bobcat (*F. rufus*)
Domestic cat (*F. catus*)
European or Scottish wild cat (*F. s. sylvestris*)
Fishing cat (*F. viverrina*)
Geoffroy's cat (*F. geoffroyi*)
Leopard cat (*F. bengalensis*)
Lynx (*F. lynx*)
Margay cat (*F. wiedi*)
Ocelot (*F. pardalis*)
Puma (*F. concolor*)
Sand cat (*F. margarita*)
Serval (*F. serval*)

● ◧ ⚘

Habitat: varies; desert, steppe, savannah, tropical and cool temperate forest.

Diet: range of small animals, ranging from insects, fish and frogs to rodents, peccaries, small deer and monkeys.

Breeding: litters of 1-6 born in den after pregnancy of 56-90 days.

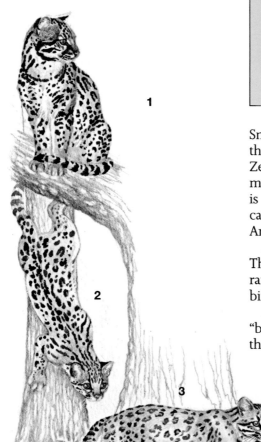

Small wild cats are found in all parts of the world except Australia and New Zealand. There are 28 species, and the most familiar is the Domestic cat. This is descended from the African wild cat, which was domesticated in Ancient Egypt about 4,000 years ago.

Almost all species are nocturnal. They hunt and stalk small animals ranging from insects, lizards and birds, to small rodents and monkeys.

All the "small cats" are similar to the "big cats" in build and behavior but they are different in other ways. They cannot roar, they eat in a crouching position, and there is a bald strip along the front of the nose. Unlike the big cats, the small wild cats tuck their front paws under the body when resting and they also wrap the tail around the body. Big cats rest with the front paws in front of them and the tail extended behind them.

CATS OF FIVE CONTINENTS
Wild cats live in many different habitats. The European wild cat makes its home in the cool forests of Northern

◀Taken by surprise, a Scottish wild cat bares its teeth in a fierce threat display. Not even a fox can get the better of it.

▼**Stalk, pounce and kill** Twelve species of small wild cat, arranged by their distribution from west (America) to east (Asia). Ocelot (**1**). Margay cat (**2**). Tiger cat (*Felis tigrinus*) (**3**). Jaguarundi (*F. yagouaroundi*) (**4**), European and African Wild cat (**5**) and (**6**). Black-footed cat (*F. nigripes*) (**7**). Sand cat (**8**). Jungle cat (*F. chaus*) (**9**). Leopard cat (**10**). Asiatic golden cat (*F. temmincki*) (**11**). Fishing cat (**12**).

Europe, while the African subspecies lives in dry, open forest and savannah. The Sand cat lives in hot, dry desert, where it preys on insects and lizards.

Some species, like the lynx, are very widespread. This lovely cat lives in forest and thick scrub from Western Europe to Siberia and in Canada and the northern United States. Also widespread, and with several subspecies, the puma can be found from southern Canada to Patagonia in South America, living in grassland, forest, steppe, desert and tropical forest. This is the largest of the North American cats and is also known as the panther, cougar or, more rarely, the Mountain lion.

Other wild cats are more special-ized or restricted in their habitats. The serval lives in African savannah country, where it preys on game birds, rodents and small deer. Never going far from water, the serval has long legs which help it run through long grass.

Even more specialized is the Fishing cat of India and South-east Asia. With its slightly webbed feet and exposed claws, the Fishing cat is well adapted for catching fish and crabs in the swamps and wet forest where it lives. It also preys on insects, birds and small mammals.

SOCIAL BEHAVIOR

Little is known about the behavior and social organization of most of the

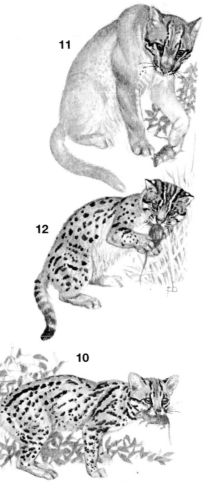

11

12

7

8

9

10

wild cats. The individuals of most species are probably territorial. As every cat-owner knows, territories are scent-marked with urine and/or droppings.

Ownership of a territory is important. It allows an individual to keep control of access to prey and mates. The size of the territory depends on the abundance of prey and the number of cats in the local area. Domestic cats living wild (feral cats) may live as solitary individuals if prey is scarce and widely scattered, or they may be in groups of 30 or more where prey is abundant.

The European wild cat also may live as a solitary individual or in a group. The territory of a solitary individual is usually larger than for the wild Domestic cat, adult males having the biggest territory.

DESIGNS FOR LIVING
The lynx and North American bobcat are similar in size and shape but are adapted for life in different kinds of habitat. The lynx lives in the cool northern forests of North America, while the bobcat extends from southern Canada to the south of Mexico.

North American winter temperatures may drop to −50°F. For this reason, the lynx has a shorter tail, which reduces heat loss, and also has a dense covering of fur on the pads of its feet. Longer legs allow it to walk through deep snow, where the bobcat may be at a disadvantage. But the bobcat's shorter legs are ideally suited for scrambling on steep rock screes and dense brush on mountainsides.

In the dense forest where the lynx preys on mice and small deer, hearing may be more important than sight. The lynx's longer ear tufts are thought to improve its hearing abilities.

As stalk-and-pounce hunters, both lynx and bobcat need to get as close to their prey as possible before making their final leap. Being able to blend in with the background is important.

The plain brownish-gray of the lynx camouflages it against the moss-covered forest trees and swamps, while the black-spotted brown coat of the bobcat blends with a background of dense brush and rock screes, where it preys mainly on cottontail rabbits.

THREAT OF EXTINCTION
Of the 28 species of small wild cat, nine are rare or threatened in some way. In some areas the danger comes from the destruction of habitats to create farmland. When this happens, wild cats may start to attack farm animals and then are killed as pests.

Another serious threat comes from the fur trade. This especially affects those small cats which have beautiful spotted fur. The Margay cat, of Central and South America, and the Leopard cat, of South-east Asia, the Philippines and Taiwan, are the victims of intense hunting pressure. Some species, such as the North American lynx and bobcat, are trapped in large numbers. But both are protected by laws that prevent too many from being killed. Geoffroy's cat from South America is another threatened species. Each year sees the slaughter of 20,000 of these lovely cats.

The ocelot may be the most beautiful of the small wild cats and it is being brought near to extinction by the fur trade. Its range extends from Arizona in the southern United States through Central and South America as far as Argentina. The ocelot is a skilled swimmer and climber and preys on reptiles, birds and small mammals. In 1975, 76,838 ocelot skins were imported into Britain alone. It takes many skins to make a single fur coat. Unfortunately, there are still enough rich and selfish people in the world to keep the trade in wild cat skins going.

► A North American bobcat pads through the snow. Winter is a lean time, when prey is scarce.

WOLVES

A long, piercing howl shatters the quiet of a northern forest. The howl grows into a chorus of many voices, and the forest valley soon echoes to the chilling sound. The leader of a wolf pack started this noise and other members joined in. The howling warns other wolf packs to keep away. There may be young cubs to protect or a kill to be guarded. For nearby farmers, the howling may mean that livestock become anxious and need to be calmed.

WOLVES Canidae (*2 species*)

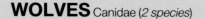

O ■ ☠

■ **Habitat:** forest, mountains, plains, desert.

◩ **Diet:** berries, small mammals, caribou, deer, moose.

◎ **Breeding:** litters of 4-7 after pregnancy of 61-63 days.

Size: head-body 3½-5ft; weight 26-176lb; males larger than females.

Color: gray to tawny-buff, varying from white in far north through red, brown to black; underside pale.

Lifespan: 8-16 years, up to 20 in captivity.

Species mentioned in text:
Domestic dog (*Canis familiaris*)
Gray wolf (*C. lupus*)
Red wolf (*C. rufus*)

▶**Wolves of the world** Red wolf (**1**). Arabian Gray wolf (**2**). Mexican Gray wolf (**3**). European Gray wolf (**4**). Tibetan Gray wolf (**5**). Gray wolf/husky cross (**6**).

Only two species of wolf survive today. The Gray wolf lives in the northern half of North America, Northern Europe and much of Asia. It is extinct in Britain and remains in Western Europe only in a few isolated areas. The decline of the Gray wolf is the result of the expansion of the human population, which led to the destruction of habitats. People have for centuries killed wolves, seeing them as a threat to farm animals. Our pet dogs are descended from wolves. The Red wolf once lived throughout south-east North America. It is now extinct in the wild except for a small group recently released in North Carolina.

LIFE IN THE PACK

Wolves are social animals and live in packs. A pack has 7 to 20 members, its size depending on the abundance of local prey. Wolves mate for life, so each pack consists of several pairs and their young.

Wolf packs have large home ranges. The smallest is about 40sq miles, while the largest may extend over 400sq miles – the size depends on the amount of food available. Within each home range, a pack has its own territory which it guards against other

▼ A wolf pack sets off in single file in search of prey. Their travels may cover an area up to 400sq miles.

▲ With lowered tail and flattened ears, a wolf greets the pack leader (1), while two cubs play (2).

packs. Wolves mark the boundaries of these areas by frequent scent-marking with urine and, less often, by howling. Wolf packs avoid each other as much as possible. When they do meet, there may be fights that result in deaths.

A strong, aggressive male leads each pack. His mate is the dominant female, and this leading pair breeds more often than the other pairs. A wolf signals its dominance by snarling and displaying its teeth, while other individuals show their acceptance of their lower position by holding their ears back and their tails between their legs. While an individual may be lower in rank than the pack leader, it can also be dominant over other members of the pack.

BREEDING AND HUNTING

Wolves breed late in the winter. A female gives birth to a litter of between 4 and 7 blind, helpless cubs in a den. After about 4 weeks the cubs leave the den. They are looked after not only by their parents, but also by "helpers" among the other members of the pack.

By hunting together, a wolf pack can run down and kill animals which would be too big for a solitary wolf. They prey on deer, caribou and antelope, as well as smaller animals. Wolves also eat berries and scavenge at rubbish heaps.

COYOTE

Pronghorn antelope graze quietly in the sagebrush country of western North America. Besides the strong scent of sagebrush, they now smell something else: danger. The danger is a pair of stalking coyotes. Pronghorns often approach moving objects, even predators. One does just that and pays for its curiosity with its life. The coyotes chase it for about 450yd and kill it quickly – they have earned another meal.

The haunting howl of the coyote is commonly heard in cowboy films, where it is used to create a feeling of night-time menace. Coyotes live in almost all of North America and extend south through Mexico as far as Costa Rica. They are medium-sized, slender members of the dog family, with a narrow muzzle, pointed ears and long legs.

They inhabit open country, occasionally mountain forest. They prey on small animals, including insects, but especially rabbits, mice and ground squirrels. Coyotes also eat fruit, carrion and larger animals such as deer and pronghorn antelope. When stalking small prey, they hunt alone, but two or more coyotes stalk larger animals.

COYOTE PACKS
Until recently, coyotes were thought to be solitary animals, but we now know that they sometimes form packs and have social lives similar to wolves. A pack usually consists of about six adults, a few juveniles of about a year old and some young. Packs of this type form because the juveniles delay leaving to form pairs of their own. Instead, they stay with their family group as "helpers" and assist in rearing the younger cubs.

Packs probably form when there is a local abundance of prey. A pack has a

COYOTE *Canis latrans*

○ ■

🔵 **Habitat:** grassland, open country, sometimes mountain forest.

◨ **Diet:** fruit, insects, carrion, small mammals, antelope, deer.

◎ **Breeding:** litters of 6 after pregnancy of 63 days.

Size: head-body 2-3ft; weight 25-33lb.

Color: grizzled buff-gray, black stripe along middle of back, black patches on forelegs and near base and tip of tail.

Lifespan: up to 14 years, 18 in captivity.

▲A lone coyote pauses to sniff the air while out hunting in Banff National Park, Alberta, Canada.

◄Announcing its presence, a coyote produces its familiar howl, which can be heard several miles away.

▼An insect or small rodent has attracted the attention of this coyote, which follows it closely.

home range of 5 to 25sq miles, and pack members scent-mark the boundaries of the home range with urine. They also use their famous howl and other sounds to signal their presence.

PARENTAL CARE

Coyotes breed from January to March. Like wolves, they mate for life, and both parents share the rearing of young. A female produces one litter each year of about six blind, helpless pups. She suckles them in a den for several weeks, but at 3 weeks, they start to eat partly digested food brought to them and regurgitated by both parents and other members of the pack.

Unlike most carnivores, the coyote is not in decline. Since the late 19th century it has expanded its range northwards and eastwards across the United States. At the same time, the Gray wolf and Red wolf were being killed in huge numbers by human settlers.

▲A coyote pack defends a kill. Three are feeding (1). The leader of the pack (2), his ears erect, tail bushy and almost horizontal and teeth exposed, threatens an intruder (3), who adopts a defensive threat posture, with tail between legs. Another male (4) backs up his dominant partner, while other intruders (5) await the outcome.

JACKALS

On the Serengeti Plain in Tanzania, African white-backed vultures swoop down to a dead zebra. But their meal is interrupted. A pair of Silverbacked jackals trot into view across the plain. They gather speed as they approach, yelping and snarling around the zebra carcass until all the birds have gone, leaving the pickings to them.

Jackals are small, slender dogs with large ears and bushy tails. Their long legs make them powerful runners. Besides carrion (the flesh of animals already dead), jackals eat almost anything, from fruit, insects and frogs to birds and small mammals.

FOUR SPECIES

There are four species of jackal. The Golden jackal is the most widespread and common. It lives in dry, short grassland in North and East Africa, South-east Europe and Southern Asia. All the other jackal species live only in Africa. The Sidestriped jackal lives in wet forest, the Silverbacked in dry, scrubby woodland.

The rarest species is the endangered Simien jackal, which inhabits the remote mountains of Ethiopia. There are only about 500 left in the wild. Jackals are often killed for their fur, and many are killed in farming areas to prevent them attacking livestock.

FAMILY "HELPERS"

Like wolves jackals mate for life. They live in family groups consisting of the

▼ Silverbacked jackals feed at a zebra carcass, having driven away vultures.

JACKALS Canidae (4 species)

● ◨ ☠

● **Habitat:** dry grassland, dry woodland, wet forest, mountains.

◼ **Diet:** fruit, small animals, carrion.

◎ **Breeding:** litters of up to 9 after pregnancy of 63 days.

Size: head-body 2-3½ft; weight 15-30lb.

Color: yellow-pale gold (Golden jackal); black and white saddle, otherwise russet (Silverbacked); reddish, with white chest and belly (Simien); gray, with white from elbow and white-tipped tail (Sidestriped).

Lifespan: up to 9 years, 16 years in captivity.

Species mentioned in text:
Golden jackal (*Canis aureus*)
Sidestriped jackal (*C. adustus*)
Silverbacked jackal (*C. mesomelas*)
Simien jackal (*C. simensis*)

▲A family group of Silverbacked jackals. A pup begs regurgitated food from the mother, while she suckles two more pups (1). A helper (2) adopts the submissive posture towards its father (3).

▼Playtime for the four species of Jackal. A Golden jackal pup plays with an adult's ear (1). Juvenile Silverbacked jackals in a tail-pulling game (2). Two nearly grown Simien jackals play a chasing game (3).A Sidestriped jackal plays with a dead mouse (4). Play is an important learning time for young hunting animals.

parents, young pups, and juveniles of about 1 year old, which sometimes for several months help their parents rear the next litter.

Each group has a territory of up to 1sq mile, which it maintains all year round. Together, the parents scent-mark the boundary of their territory with urine. Both parents and juveniles hunt together. This way they can run down larger prey which would be too big for a single jackal.

Golden and Silverbacked jackals mate in October, at the end of the dry season. Pups are born in a den during December and January, the wet season, when food is plentiful. Females produce up to nine pups, which remain in the den for up to 3 weeks. Their mother spends nearly all her time with them, but the juveniles guard the pups when the mother is away hunting.

Jackal pairs which have a team of juvenile "helpers" can raise more young than those without helpers. The advantage to helpers is that they become familiar with the home range of their parents, part of which they may inherit.

FOXES

It is a warm, moist evening. The Red fox is quietly padding across the grassy meadow, ears pricked for the slightest sound. Then he hears it, the rasping of an earthworm's bristles on the grass. Finding the exact source of the sound, he poises, before plunging his muzzle into the grass to grab the worm. The worm, though, still has its tail in its burrow. So the fox pulls it taut and tugs until the worm comes free.

FOXES Canidae (21 species)

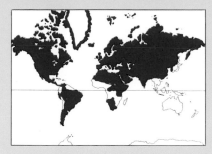

● ▢ ☠

Habitat: general, including urban.

Diet: small mammals, birds, insects, eggs, fish, fruits, berries, carrion.

Breeding: up to 8 offspring after pregnancy of 3 months (Red fox).

Size: smallest (Fennec fox): from head-body 10in, tail 7in, weight 1¼lb; largest (Small-eared dog): up to head-body 40in, tail 14in, weight 20lb.

Color: gray to reddish-brown coat, sometimes white, silver, cream or black.

Lifespan: up to 6 years.

Species mentioned in text:
Arctic fox (*Alopex lagopus*)
Argentine gray fox (*Dusicyon griseus*)
Azara's fox (*D. gymnocercus*)
Bat-eared fox (*Otocyon megalotis*)
Colpeo fox (*Dusicyon culpaeus*)
Crab-eating fox (*D. thous*)
Fennec fox (*Vulpes zerda*)
Gray or Tree fox (*V. cinereoargenteus*)
Indian fox (*V. bengalensis*)
Kit or Swift fox (*V. velox*)
Red fox (*V. vulpes*)
Small-eared dog (*Dusicyon microtis*)

◄ A Gray fox keeps watch from its vantage point up in a tree. It has the white throat typical of many species of fox.

▼ **Eight vulpine species** Foxes of the genus *Vulpes* shown dashing after and swiping at a bird. Gray fox (**1**). Swift fox (*Vulpes velox*) (**2**). Cape fox (*V. chama*) (**3**). Fennec fox (**4**). Rüppell's fox (*V. rüppelli*) (**5**). Blanford's fox (*V. cana*) (**6**). Indian fox (**7**). Corsac fox (*V. corsac*) (**8**).

Foxes, like dogs, belong to the family of canids. Compared with dogs, they have a flattened skull, a pointed muzzle and a long bushy tail. Their triangular ears are fairly big and stand erect. The tip of the tail is often a different color from the rest of the coat, usually black or white. Several species, including the Red fox, have a white chin.

THE TYPICAL FOX

The Red fox is often thought of as the "typical" fox because it is found so widely. But, with a head-to-tail length of over 44in and weight up to 13lb, it

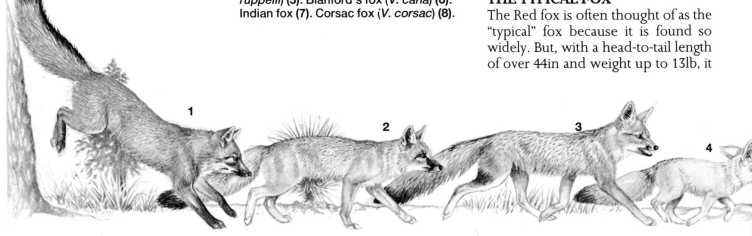

is bigger than its relatives in the genus *Vulpes*. More typical is the slighter-built Indian fox, which inhabits the open forest, scrub and steppe land of India, Pakistan and Nepal. This fox has a sandy-brown coat, with darker legs and tail.

The Gray fox of North and South America can grow nearly as big as the Red fox. It is also known as the Tree fox because it has the habit of climbing trees. It often sleeps in trees.

MOST ADAPTABLE
The Red fox is the most widespread and most successful of all the fox species. It is found from the far north of North America, Europe and Asia, south to the deserts of Central America and North Africa. It can adapt to a variety of climates from the frigid cold of the Arctic to the searing heat of the desert. It can also adapt to life in the city.

The Red fox is also very adaptable when it comes to food. It will eat almost any food that is available, not only small mammals, birds, eggs, worms, rabbits, but also in season fruits such as blackberries and apples and even rose hips. Fruits can form as much as 90 per cent of its diet.

▲ Two male North American Red foxes battling in the woods on a fall day. It is a test of strength to see which one will be dominant.

5

6

7

8

HUGE EARS

One species of fox is immediately recognizable by its huge ears. It is the well-named Bat-eared fox of Africa. This animal is also notable because of its diet, which is mainly of insects. Termites and dung beetles and their larvae are among its favorite foods. Its hearing is so sensitive that it can detect the sound of larvae gnawing their way out of dung balls.

The cream-colored Fennec fox, smallest of all the foxes, also has very large ears. It lives in the sandy deserts of Africa and Arabia. Its large ears not only enable it to locate its prey. They also act as "radiators" to get rid of excess heat, helping the animal stay cool. Other adaptations to desert life include furry feet, which help the Fennec keep its footing in the sand.

The ears of the Arctic fox, by contrast, are small and rounded. The muzzle is shorter than in the typical fox. These are ways in which this fox has adapted to living in a bitterly cold climate. The very thick underfur of its winter coat insulates the fox. This fur conserves body heat until the temperature falls to below minus 94°F. Only then does the fox start to shiver!

▲ The coat color of the Arctic fox can vary from animal to animal and from season to season. A polar form has an all-white coat in winter (1), which becomes darker in summer (2). Other foxes may have a steel-gray or brown (3) coat in winter. An Arctic fox vixen in her brown summer coat is seen here with younger brothers and sisters, each of which has a different colored summer coat (4).

▼ South American foxes of the genus *Dusicyon*. Small-eared dog (1). Colpeo fox (2). Argentine gray fox (3). Azara's fox (4). Crab-eating fox (5).

IN THE DEN

Foxes breed once a year. The vixen, or female fox, gives birth to her young in a den, which is often a burrow or sometimes a hollow in a tree or a rock crevice. The young pups, or cubs, are helpless at first and remain in the den suckling their mother for several weeks. By the age of 6 months they can hunt by themselves.

Most species of foxes lead solitary lives, coming together only at mating time. But members of some species may share their dens with other foxes. Red and Arctic foxes sometimes do this. A typical group would include one male and several vixens, probably all related. They feed in different parts of the group's territory, but their paths may cross many times each night.

◄The Kit fox lives on the prairies of North America. Now classed as one of the Swift foxes, it is smaller and has larger ears than the other North American foxes.

►Like other members of the dog family, foxes mark their territory with urine, especially the places they visit often.

WIDE-SCALE HUNTING

The skins of foxes have long been prized for making fur coats and wraps. Foxes are still hunted on a large scale to satisfy the demand. In North America alone, more than half a million Red, Gray and Arctic foxes are killed each year for the fur trade. The white and blue furs of the Arctic fox are especially in demand. Several states now limit fur licenses severely and restrict fox hunting to just a few weeks each year.

Large numbers of foxes are also killed where they have become pests, and also to combat the spread of the deadly virus rabies. Other foxes are hunted for sport. But despite destruction of their habitat, few species of fox are in danger of extinction. Perhaps the most at risk are three South American species, the Argentine gray fox, the Colpeo fox and the Small-eared dog of the tropical forests. This is the largest of all the foxes.

▼Two alert Bat-eared foxes, unusual among the foxes in living mainly on insects. They use their large ears to detect insect prey.

WILD DOGS

A herd of wildebeest graze in the cool of an East African dawn. Suddenly, they smell danger in the air and are off, with a pack of African wild dogs hard at their heels. Two of the dogs pick out an old wildebeest which is lagging behind. With a burst of speed, one has the wildebeest by the tail.

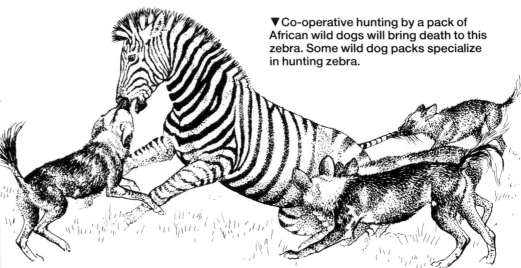

▼Co-operative hunting by a pack of African wild dogs will bring death to this zebra. Some wild dog packs specialize in hunting zebra.

WILD DOGS Canidae
(6 species)

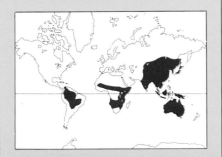

 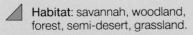

▲ **Habitat:** savannah, woodland, forest, semi-desert, grassland.

■ **Diet:** berries, insects, reptiles, mammals from rodents up to wildebeest and zebra.

◯ **Breeding:** litters of 2-5 after pregnancy of 60-73 days.

Size: head-body 20-42in; weight 11-60lb.

Color: from white and yellow blotching on dark background, through tawny and foxy red to grizzled-gray.

Lifespan: where known, 10-14 years.

Species mentioned in text:
African wild dog (*Lycaon pictus*)
Bush dog (*Speothos venaticus*)
Dhole or Asian wild dog (*Cuon alpinus*)
Dingo (*Canis dingo*)
Maned wolf (*Chrysocyon brachyurus*)
Raccoon dog (*Nyctereutes procyonoides*)

Wild dogs live in packs of 2 to 20 adult animals, but the most common number is 7 or 8. Each pack has a home range, of up to 600sq miles. They criss-cross their range in search of prey and only have a fixed base when there are young pups.

African wild dogs prey mainly on impala, gazelles and sometimes smaller animals such as rats. The pack travels, rests and hunts together. They kill their prey quickly. If there are any pups, they are allowed to eat first.

UNUSUAL SOCIAL LIFE
The African wild dog lives in savannah woodland and desert. There are probably no more than 10,000 left, and the species is regarded as endangered.

The social life of the African wild dog is the opposite of other predators. Males remain in the pack they were born into and so are related to each

▼Two female African wild dogs in a tug of war over a pup in a battle for breeding dominance. Pups are often killed by this.

▲Sizing each other up, two Maned wolves circle each other, the one on the right arching its back to make itself look larger.

▼A dingo, Australia's wild dog, pulls back as a lizard it is attacking tries to defend itself.

blind pups weighing about 1lb. Pups are born in a den, where both mother and young feed on regurgitated food from other members of the pack.

FOREST AND GRASSLAND DOGS

The dhole or Asian wild dog has a similar family life to the African wild dog. It lives in forests, from India to China and South-east Asia, where it eats berries, insects, lizards, rodents and deer.

The wild dog of Australia is the dingo. Like the Domestic dog, this species is descended from the wolf and was probably brought to Australia by Aboriginal peoples about 20,000 years ago.

In Eastern Asia the Raccoon dog is most un-dog-like in appearance. As its name suggests, it looks more like a raccoon. A forest dweller, this small dog has been introduced into parts of Europe.

The Maned wolf of Central South America looks rather like a fox on stilts. Its long legs help it to move through long grass. The little-known Bush dog of South America has short, squat legs, tiny ears and a blunt face.

other. The young females – usually more aggressive than the males – leave the pack as a group of litter-mates to join the males of another pack.

The African wild dog mates once a year, when food is plentiful. Usually only the dominant male and female of the pack mate. Females give birth to

POLAR BEAR

A hole appears in a smooth bank of deep snow, high in the Arctic. Out of the hole peeps a black nose. The hole gets larger and the black nose is followed by a broad, massive white head. Eventually the huge bulk of a female Polar bear looms out of the hole and moves slowly down the bank. She stops and waits as two smaller heads appear at the hole. Her cubs blink at their first sight of the Sun as they leave the den where they were born.

Polar bears are the largest living four-legged carnivores. They survive in one of the harshest areas of the world, braving the freezing cold of the high Arctic. Much of their time is spent on the pack-ice, far from land.

Everything about Polar bears is geared for survival in extreme cold. They are well insulated by a thick fur coat and a layer of fat – only the nose and pads of the feet are without fur. Small ears and a very short tail also prevent the loss of body heat.

PATIENT HUNTERS

Polar bears are excellent swimmers and can swim for hours through icy-cold water. Their feet are slightly webbed, and each foot has five long, curved claws. The claws help them to grip not only the slippery ice but also their prey. And the white coat camouflages a Polar bear against the snow, allowing it to sneak up unnoticed on basking seals.

Polar bears mostly hunt seals, especially the Ringed seal. For most of the year they hunt by waiting patiently for seals to appear at their breathing-holes in the ice. In April and May they break into the dens of Ringed seals, killing the mothers and the pups.

Polar bears may even attack and kill small whales and walruses. They also

POLAR BEAR
Ursus maritimus

● ◻ ☠

■ **Habitat:** sea ice and waters, coasts and islands.

■ **Diet:** seals, small whales, walrus, carrion, plants.

◎ **Breeding:** litters of 1-3 after pregnancy of about 8 months.

Size: head-body 6-7½ft, weight 385-660lb female; head-body 7½-9ft, weight 770-1400lb male.

Color: white, or yellowish.

Lifespan: 20-25 years.

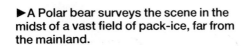

▶A Polar bear surveys the scene in the midst of a vast field of pack-ice, far from the mainland.

eat carrion and, in summer, some plant material. Polar bears are mostly solitary animals, though up to 30 may gather at a good food source such as a dead whale.

A male Polar bear finds a female on heat by following her smell. A single male may have a large home area which includes the home areas of several females. Breeding begins when Polar bears are 5 years old. They mate in April, May and June.

WINTER IN THE DEN

In November and December pregnant females dig dens in the snow. Each female remains in her den until March, and it is here that she gives birth to a litter of up to three cubs, each weighing 21 to 25 ounces.

Polar-bear milk is about one-third fat, which helps the cubs to keep their body temperature up through the permanent night of the Arctic winter. Although the female Polar bear does not feed while she is in the den, she is not in true hibernation.

The cubs leave the den with their mother when they are about 3 months old and weigh 17½ to 26½lb.

▲A cub follows its mother across the ice. It will remain with her until it is about 28 months old.

▼Polar bears scavenge at a rubbish heap in Alaska. This happens when towns are built on bear migration routes.

GRIZZLY (BROWN) BEAR

It is spawning time for the Pacific salmon. As the large fish labor upstream along a Canadian river, a Grizzly bear and her cubs come tumbling down the bank into the shallows, where the salmon are leaping. With her claws and teeth, the female Grizzly catches fish after fish and neatly strips the flesh from both sides.

Grizzly bears are famous for their strength and speed. A Grizzly can bite through a steel bolt ½in thick. Despite its lumbering weight of nearly half a ton, it can charge at 30mph.

Grizzly bears are forest animals and today are found in north-west North America and Europe and northern Asia. They are often called Brown bears and are rare, with only a few small isolated populations remaining.

Grizzlies eat mainly plants, especially young leaves and berries – they use their strong claws to dig up tubers and roots. They also eat insect grubs, rats and mice, salmon, trout and young deer. Sometimes Grizzlies will attack farm animals.

SPRING COURTSHIP

Male Grizzlies are solitary animals. Each adult male has a home range of up to 420sq miles, which he defends fiercely against other males. They often fight to the death. Females also have territories, which they may share with a few young daughters. Their territories are smaller than those of males, up to 75sq miles. Females, too, defend their territories against other females, so that they can enjoy exclusive access to food.

In spring, male Grizzlies seek out females. After a short courtship of 2 to 15 days they mate in May and June. As with all bears, the fertilized egg does not implant in the womb until October or November. At this time, the

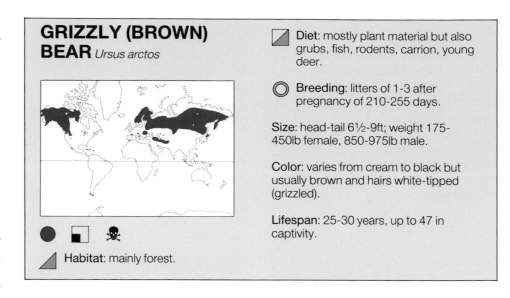

GRIZZLY (BROWN) BEAR *Ursus arctos*

▨ **Diet:** mostly plant material but also grubs, fish, rodents, carrion, young deer.

◯ **Breeding:** litters of 1-3 after pregnancy of 210-255 days.

Size: head-tail 6½-9ft; weight 175-450lb female, 850-975lb male.

Color: varies from cream to black but usually brown and hairs white-tipped (grizzled).

Lifespan: 25-30 years, up to 47 in captivity.

◉ ▣ ☠

◣ **Habitat:** mainly forest.

female either digs out her own den or finds a natural cave or a hollow tree. She remains there throughout the winter, not feeding, but relying on her store of body fat.

She gives birth to two or three naked, helpless cubs, each weighing only 12½ to 14 ounces. They remain in the den until April, May or June and stay with their mother for up to about 4½ years.

FEW SAFE AREAS

Grizzlies are now endangered in many parts of North America. They are sometimes shot when they feed at rubbish heaps and have been wiped out in many areas.

Despite their fierce reputation, Grizzlies rarely attack people. When a bear does attack, it is sometimes because with its poor eyesight it mistakes a person for another bear.

▼ Grizzlies love water and spend much time in and around salmon rivers bathing, frolicking and feeding.

► A bear hug: two young male Grizzlies in a play fight, which is good practice for adult fighting.

AMERICAN BLACK BEAR

A black bear forages among the bushes. Her two cubs clamber up a nearby tree and slither down the trunk. They jump on one another and wrestle, rolling over on the ground. They bump into their mother, and she cuffs them with her paw. But they carry on playing. The mother bear puts up with their games. She is too busy eating to join in. She is thin after having the cubs, and she is making good use of the late summer to feed on berries.

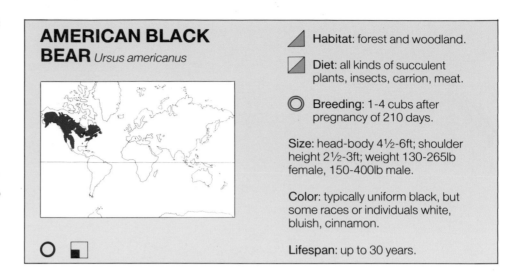

AMERICAN BLACK BEAR *Ursus americanus*

Habitat: forest and woodland.

Diet: all kinds of succulent plants, insects, carrion, meat.

Breeding: 1-4 cubs after pregnancy of 210 days.

Size: head-body 4½-6ft; shoulder height 2½-3ft; weight 130-265lb female, 150-400lb male.

Color: typically uniform black, but some races or individuals white, bluish, cinnamon.

Lifespan: up to 30 years.

▼Apart from an occasional dash to catch prey, the pace of life for an American black bear is slow. For many, in the northern parts of their range, food is scarce and slow-growing. Bears are often seen scratching the ground for food, like the one shown here. A female black bear may wander over 40sq miles in a year to find enough food.

The American black bear once roamed over most of the woods of North America from central Mexico to Canada. Where people have built towns or cleared the woods, the bears have vanished with their habitat. But where there are forests, American black bears can still be found. They do not often go into open country.

In spite of their name and usual color, there are several varieties of black bear. Near the west coast in British Columbia there are even pure white "black bears," as well as brown and bluish forms. In the east pure black bears are usual. This is also where the largest bears come from, the biggest on record being a male that reached 600lb and almost 7¾ft from nose to the tip of its stubby tail.

▼Black bears like meat if they can catch it, or will feed on carrion. This one has found a cow killed in a storm.

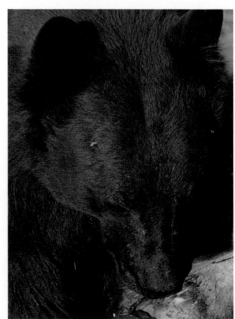

SWEET TOOTH

An American black bear needs up to 18lb of food a day. It feeds on bulbs, tubers, young shoots, and nuts and berries in season. It also kills animals from the size of mice to young deer if it has the chance. It digs up insect grubs and is fond of honey and sweet things. The black bear is a good climber and sometimes goes up trees in search of food.

Black bears mate in the summer. In the winter most den up, sometimes for as long as 7 months, so they spend more than half their lives asleep. While asleep during the winter the heartbeat slows down, and the body's temperature drops a little. This cuts down the bear's use of energy.

It is in the den, in January or February, that the cubs are born. Each cub is naked and weighs only 9 ounces. They stay denned with the mother through the cold weather until the spring and suckle until the late summer. They may stay with the mother as long as 2½ years.

When the cubs first emerge from the den they weigh about 4½lb. The mother is very protective. Her alarm grunt may send the youngsters scrambling up a tree out of reach of enemies. Accompanying their mother, the cubs learn how to dig out all kinds of delicacies.

Bears do not begin to breed until they are from 3 to 5 years old. Even in good conditions a female has cubs only once every 2 years.

BEGGING BEARS

Black bear numbers have declined as people have settled across North America, but there are more left than there are Brown (Grizzly) bears. They usually avoid people, but when hungry they may scavenge for scraps. In National Parks, where they are not molested, some bears get into the habit of begging for food beside roads. They can become dangerous if not given what they want.

◄Black bears scavenge at waste dumps or garbage cans. This habit brings them into conflict with people.

▼An American black bear kills an unwary beaver and pulls it ashore to eat. Such meals supplement a mainly vegetable diet.

SMALL BEARS

High in the tree a buzz sounds from a crack in the bark where wild bees have a nest. On the ground below a Sun bear pauses and listens. Then it shins up the tree trunk straight to the nest. It scratches at the bark with its big claws until it has ripped open the nest. Pulling out the honeycomb, it eats honey and bee grubs.

Three of the four smaller bears live in the south of Asia. The fourth, the Spectacled bear, is found in the Andes in northern South America. Although they are small as bears go, all can weigh as much as an adult human. Males are bigger than females and (except the Sun bear) may be much heavier than a man.

SMALLEST BEAR

The Sun bear, from the tropics of South-east Asia, is the smallest bear and has the shortest and sleekest coat. Like the other Asian small bears it has a light colored mark on its chest. The Sun bear is a forest dweller. It is light, and its big curved claws and large naked soles can give a good grip as it climbs.

The Sun bear spends much of its time in the trees. Active at night, it often sunbathes during the day, and this gives the animal its name. It eats all kinds of foods, from fruits and young tips of palm-trees to insects, small mammals and birds.

▼Performing bears, once common, are a rare sight today. Bears like this Sloth bear can look "human" standing on their hind legs.

SMALL BEARS Ursidae
(4 species)

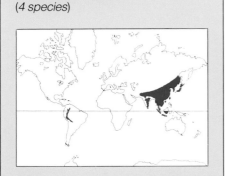

● ◰ ☠

 Habitat: forest, some grassland and scrub.

 Diet: omnivorous.

○ Breeding: 1-3 (usually 2) cubs after pregnancy of 7-8 months (3½ months for Sun bear).

Size: smallest (Sun bear): head-body 3½-4½ft, weight 60-150lb; largest (Sloth bear): head-body 5-6½ft, weight 200-300lb.

Color: black, with lighter markings on muzzle and chest.

Lifespan: up to 30 years.

Species mentioned in text:
Asian black bear (*Selenarctos thibetanus*)
Sloth bear (*Melursus ursinus*)
Spectacled bear (*Tremarctos ornatus*)
Sun bear (*Helarctos malayanus*)

▲ **The four smallest species of bear** The Sloth bear (1) makes good use of its long curved claws and flexible snout to forage, on the ground for termites and grubs, or in trees. A Sun bear (2) licks termites from its paw after scooping them out of the mound it has broken open. The Spectacled bear (3) is a good climber. It eats tree fruits, palm frond bases and other succulent plant food. The Asian black bear (4) sometimes kills large animals or feeds on carrion like this deer carcass.

SUCKING UP INSECTS

The Sloth bear has shaggy fur and long claws. It eats many things, but prefers insects. It uses its claws to break open termite mounds, then sucks up the inhabitants. Its flexible lips can be formed into a tube, and there is a gap in the front teeth. The nostrils can be closed to prevent insects and dust getting in.

The Sloth bear blows dust from its prey before sucking them up. It is mainly nocturnal and likes to live in forested areas.

Further north lives the Asian black bear. This bear is omnivorous, feeding largely on plants, but also eating animals. At times it has made itself unpopular with people by killing domestic animals or eating crops. It usually avoids humans, but can be aggressive if cornered. Although it is called "black bear," some individuals are brown or reddish.

MOUNTAIN BEAR

The Spectacled bear lives mainly in damp forests, but also goes into mountain grassland and lowland scrub deserts. It can be found at any height up to 14,000ft in the Andes. The eye markings that give this bear its name are very variable. The spectacled bear is solitary and rather shy, so it is difficult to observe and study.

This bear is mainly a plant-eater but can catch young deer and guanacos. Like the other bears in the tropics it seems to be active all year and may produce young at any time. Black bears in colder regions may retire to their den for the winter.

COATIS

The tree is full of coatis. On every branch one or two are resting, scratching or grooming one another. From a nest of sticks on the next tree a female coati emerges. Soon she is down with the troop, bringing three youngsters just ready to leave the nest. The troop members look and sniff at the new arrivals, then carry on with what they were doing before. One female, especially friendly with the mother, stays and grooms the young coatis.

Little is known about two of the species of coati. One of these lives only on Cozumel Island off the eastern coast of Mexico. Another lives in the mountain forests of Ecuador and Colombia in South America. The best known species are the White-nosed coati, which lives in the southern USA, and as far south as Ecuador, and the Ring-tailed coati, which lives in tropical forests east of the Andes mountains. Coatis are good climbers and are mainly active during the day.

COMPANIONS AND LONERS

Female coatis are very sociable animals. They usually live in troops of 5 to 12 animals. Sometimes there can

COATIS Procyonidae
(*4 species*)

● ■

◢ Habitat: tropical lowlands, dry mountain forest, forest edges, some grassland.

◩ Diet: insects, other small animals up to size of lizards, fruits.

◎ Breeding: 2-5 young after pregnancy of 77 days.

Size: head-tail 2¼-4¼ft, over half tail, weight 8-12¼lb.

Color: reddish, grayish or olive brown with some contrasting markings.

Lifespan: up to 10 years.

Species mentioned in text:
Ring-tailed coati (*Nasua nasua*)
White-nosed coati (*N. narica*)

◄Coati "friends" will often groom one another (1). Some group members stay on watch (2). Mothers groom their young (3), and the young often play together (4).

▼A White-nosed coati drinks with its tail up. The tail is often held high when searching for food.

be as many as 40. A troop lives in an area ½ mile across. Females may form close friendships. They groom one another and look after each other's young. At night the troop members sleep together in the trees.

Adult males, on the other hand, are usually solitary. They join the troop during the mating period, but at other times of the year they live alone, although their home range overlaps with a troop. Because they behave so differently, and are more heavily built, these solitary males were once thought to be a different species called coatimundi.

FOLLOWING THEIR NOSES

A troop of coatis forage for food by pushing their noses into every likely log and piece of leaf litter or under stones. Their sense of smell is good, and the coati's nose is a good digging organ, as are the claws.

◄A Ring-tailed coati shows its strong front claws and the long mobile snout it uses to dig and sniff out prey.

Coatis find a whole range of small animals to eat, including beetles, grubs, ants, termites, spiders, scorpions, centipedes and land crabs. They also catch frogs, lizards and mice and will eat the eggs of birds and reptiles. They also enjoy fruit.

TELLING TAILS

A coati has a long tail which acts as a balancing rod while it is climbing. The ring markings on the tail make it a good flag to show where a coati is, what it is doing and what kind of mood it is in. Coatis also make high-pitched twittering noises that help them keep in touch with one another while they move through the forests.

◄A White-nosed coati finds an insect among the leaves. The curiosity of these animals often helps them find food.

RACCOONS

Late in the evening, at the edge of a marsh sits a raccoon. It reaches down with its hands to feel and grope in the water among the mud and plant roots. The raccoon does not seem to watch what its hands are doing. Suddenly it pulls out a hand, grasping a small frog. Dunking its catch back in the water, it washes the frog backwards and forwards, before finally eating it.

Raccoons are American relatives of the pandas of south-east Asia. Four species are each found on a single island. Two species are widespread.

The Common raccoon lives as far north as southern Canada and as far south as Central America. The Crab-eating raccoon is found in Central America and northern South America. Raccoons eat a wide range of foods, foraging in trees, on the ground and also in water.

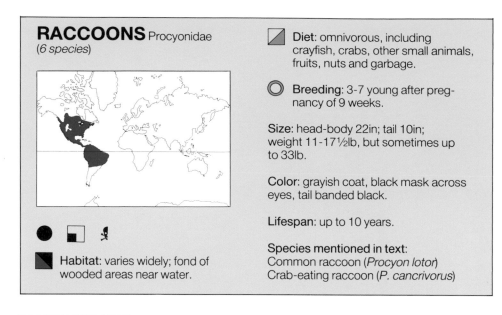

RACCOONS Procyonidae
(*6 species*)

● ■ ☠

Habitat: varies widely; fond of wooded areas near water.

Diet: omnivorous, including crayfish, crabs, other small animals, fruits, nuts and garbage.

Breeding: 3-7 young after pregnancy of 9 weeks.

Size: head-body 22in; tail 10in; weight 11-17½lb, but sometimes up to 33lb.

Color: grayish coat, black mask across eyes, tail banded black.

Lifespan: up to 10 years.

Species mentioned in text:
Common raccoon (*Procyon lotor*)
Crab-eating raccoon (*P. cancrivorus*)

MOBILE HANDS
A raccoon's hands are very mobile and have a good sense of touch. It uses them to explore food, especially in water. This behavior is so ingrained that a raccoon in captivity may take food, place it in water, and then retrieve it. This has led to the myth that raccoons "wash" their food.

LIVING WITH PEOPLE
Raccoons are fond of crayfish, but they will eat foods ranging from mice and worms to birds and their eggs. They also eat fruit and corn and sometimes raid crops, which makes them unpopular with farmers. They are curious and will investigate all kinds of places and sources of food.

▲ The Crab-eating raccoon is a good climber and has a long tail. It also spends time in and near water.

► A Common raccoon rests at the entrance to its den. Raccoons are active mostly during the night.

Raccoons sometimes make dens in barns and sheds, living close to people and making a living on their left-overs. This is probably one reason why the Common raccoon has been able to spread northwards in recent years. Unfortunately, it is an important carrier of the disease rabies, so its presence is not always welcome.

▲Raccoons live alone, but many may gather where food is plentiful. These are begging for scraps at a roadside.

▼A female Common raccoon suckles her babies in her tree den. Young raccoons stay with their mother for a year.

▲A Common raccoon raids a trash can. Raccoons have learned to live in areas where there are many humans.

WINTER DENS

In the northern part of its range the Common raccoon puts on fat in the fall. It stays in a den for much of the winter, although it does not truly hibernate. Dens may be in hollow trees or below ground. A family of raccoons may den together, even though the young go their own way to find food. As many as 23 raccoons have been found squeezed together in a winter den.

Common raccoons are born in the spring. A mother produces her litter in the security of a tree den. The young are tiny and undeveloped, only about 2½ ounces in weight. It is 7 weeks or more before they can leave the den unaided, and not till 10 weeks old do they follow their mother.

Raccoons begin to breed when they are 1 or 2 years old. Males grow somewhat bigger than females.

WEASELS AND POLECATS

A weasel crosses the open ground between two hedges. Its movements are so swift and flowing that anyone watching would hardly realize that an animal had passed across. It dives into the hedgerow and works its way along, poking its head into every hole and crevice in case there is something to eat. A movement in the leaves alerts it. A mouse is there. Like lightning, the weasel goes in pursuit. The mouse dives down its burrow, but there is no escape. The weasel goes down after it. At the bottom of the burrow it bites savagely at the back of the mouse's head. The weasel eats the brain and head, then the rest of its victim.

There are some 67 species in the mustelid family, which includes such animals as badgers and otters. There are 21 relatively small members of the family that make up the species of weasels, stoats, polecats and mink. They are found over most of the world in many habitats.

FEROCIOUS HUNTERS

All of this group are ferocious hunters. They differ from most carnivores in that they are able to kill, single-handed, animals bigger than themselves. They are intelligent and energetic. The smallest weasels are, if anything, even more lively and ferocious than the larger ones.

Weasels have long slim bodies and short legs. The head is narrow too, and the animal can get through a tiny space. A European common weasel is said to be able to squeeze through a wedding ring.

The European common weasel feeds on mice, voles and any other small animals it can capture. It is

▲A stoat in its summer coat, chestnut with a light bib. In cold areas stoats change to white (ermine) in winter.

►The long low shape of weasels and polecats allows them to squeeze through small openings after prey. Here a European polecat hunts down a rabbit burrow. This is the species from which the domestic ferret was bred.

WEASELS AND POLECATS Mustelidae; sub-family Mustelinae (*21 species*)

Size: smallest (Least weasel): head-body 5in, tail 1in, weight 1 ounce; largest (grison): head-body 19-22in, tail 6½in, weight up to 7lb.

Color: usually brown above, light below; others with black and white markings or other bold patterns.

Lifespan: up to 10 years, usually much less in wild.

Species mentioned in text:
Black-footed ferret (*Mustela nigripes*)
European common weasel (*M. nivalis nivalis*)
European polecat (*M. putorius putorius*)
Grison (*Galictis vittata*)
Least weasel (*Mustela nivalis rixosa*)
Long-tailed weasel (*M. frenata*)
Stoat or ermine (*M. erminea*)

● ◨ ⚐

■ **Habitat:** all types.

■ **Diet:** rodents, rabbits, birds and other animals.

◎ **Breeding:** 2-13 young after pregnancy of 5-7 weeks, plus delay in some species.

capable of killing rabbits, although these are more usually prey for the larger stoat. Polecats eat rodents, worms, carrion and also birds, which is why gamekeepers dislike them.

Most of this group hunt on or under the ground, but mink are also at home in the water. They have partly webbed feet and catch fish as well as land animals.

Most of the weasel group feed entirely on other animals, content to eat whatever they can catch. The Black-footed ferret of the American prairies, though, is a specialist, feeding just on prairie dogs. Because these rodents have been exterminated as a pest in many places, the ferret is endangered.

COAT FOR ALL SEASONS

Some of the weasel family change their coat for the winter months. The Least weasel in North America and the European common weasel in Northern Europe turn white in winter. Some other Asian species also get a lighter coat. The stoat can also change, except for the tip of the tail, which always remains black.

In the snows of the northern winter these animals are as well camouflaged from their prey as they are with a brown coat in summer.

ROUGH COURTSHIP

Weasel species are usually solitary. At most times of the year even a male and a female will try not to come close.

Males in most species are larger than females. In small species, such as

▶**Species of the south** The North African banded weasel (*Poecilictis libyca*) **(1)**. The African striped weasel (*Poecilogale albinucha*) **(2)**. The Marbled polecat (*Vormela peregusna*) **(3)** lives on the Asian steppes. The zorilla (*Ictonyx striatus*) **(4)** of Africa. The Little grison (*Galictis cuja*) **(5)** of South America and the European polecat **(6)**. The Patagonian weasel (*Lyncodon patagonicus*) **(7)** and the Black-footed ferret **(8)** are shown hunting.

the European common weasel, the size difference is very great, and males are twice as heavy as females. An advantage may be that the sexes hunt different prey and do not compete for food.

Courtship is rough, with the male grabbing the female by the scruff of the neck before mating. Afterwards the male takes no further interest in female or young.

UNUSUAL PREGNANCY

After mating, the fertilized egg begins to develop, but in many species of the weasel family the development is interrupted. The embryo remains dormant for a while, before attaching to the inside of the mother's womb and developing through to birth.

Not all weasels have this "delayed implantation," but it occurs in the stoat and Long-tailed weasel, as well as in some of the related badgers, otters and martens. It may have come about to help birth take place at a season when there will be plenty of food.

Most of the weasel group are quite small animals, and one would expect the females to be pregnant for only a few weeks. But because of delayed implantation the time gap between mating and giving birth can be up to as much as one year.

DEVOTED MOTHERS

Female weasels are devoted mothers. They give birth to blind and thinly furred babies in a secure nest. Even after weaning the youngsters stay with their mother and learn to hunt. The European common weasel can have up to three litters a year and may be capable of breeding at only about 4 months of age.

▶The European polecat has distinctive face markings. In England it was hunted so much that it became extinct.

MARTENS

From high in a pine tree a Pine marten looks out, searching for prey. It surprises a squirrel sunning itself near by. The marten leaps at it. The squirrel is already running out along a branch and jumps to the next tree. The marten follows, running through the tree-tops. It gains on the squirrel. But then the squirrel runs out along some thin twigs and jumps across a clearing. The marten cannot follow.

Most martens live in Europe and Asia. Two species, the American marten and the fisher, live in North America. Martens spend some time on the ground, but all can climb well. The long bushy tail helps with balance. Their large paws have strong claws and are furred below, giving a good grip on tree trunks and branches.

AGILE HUNTERS
Martens make leaping from branch to branch look easy. They are fast and agile. They look into likely hiding-places for prey. If they see prey they make a quick rush and kill it with a bite to the back of the neck. Sometimes they chase prey through the trees. Squirrels are the main prey for some.

The fisher, being less agile in trees, spends more time on the ground than

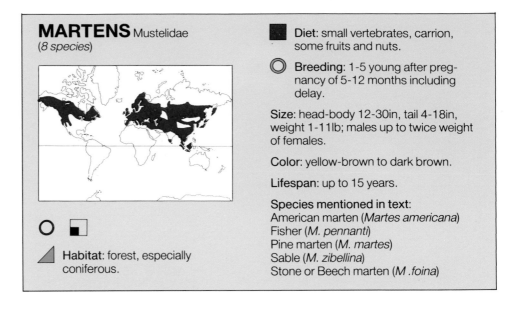

MARTENS Mustelidae
(*8 species*)

○ ■

△ Habitat: forest, especially coniferous.

■ Diet: small vertebrates, carrion, some fruits and nuts.

○ Breeding: 1-5 young after pregnancy of 5-12 months including delay.

Size: head-body 12-30in, tail 4-18in, weight 1-11lb; males up to twice weight of females.

Color: yellow-brown to dark brown.

Lifespan: up to 15 years.

Species mentioned in text:
American marten (*Martes americana*)
Fisher (*M. pennanti*)
Pine marten (*M. martes*)
Sable (*M. zibellina*)
Stone or Beech marten (*M .foina*)

most martens. It is one of the few animals that can catch a porcupine. It attacks and worries it around the head, avoiding the quilled back and tail. When the porcupine tires, the fisher rushes in, bowling it over and attacking and eating the soft belly with its sharp teeth and strong jaws.

BEHAVIOR AND BREEDING
Martens are usually solitary. They mark their pathways, in trees or on the ground, with scent from the anal glands and urine.

Young are usually born in early spring. Babies are blind and deaf and have little fur, but by 2 months old they are weaned. By 4 months they

can kill prey, and soon after this they leave their mother and live alone.

Martens are active by both day and night. Even in the bitter cold of the northern winter they leave their dens to hunt – they have very warm coats. Martens, in particular sables and fishers, are prized for their fur. Huge numbers used to be trapped. Some protection is now given, and the species are not in danger.

▶ The Stone marten lives across most of Europe. It often lives on rocky ground, and will also enter built-up areas.

▼ The Stone marten (1) is more heavily built than the Pine marten (2) and spends less time hunting in the trees.

WOLVERINE

On the spring snow a large bloodstain shows where a wolverine has killed a deer. The wolverine has fed well. Now it is breaking up what remains of the carcass with its strong jaws. It drags a deer leg away to some rocks on a hillside and rams it deep into a crevice. Marking the place with scent, the wolverine then returns to the carcass. It takes the next piece to another hiding place. The hidden food will be useful when the wolverine is hungry.

WOLVERINE *Gulo gulo*

○   ◧

◺ Habitat: Arctic and sub-Arctic tundra and taiga.

■ Diet: deer, small mammals, birds, carrion and some plants.

◎ Breeding: 1-4 kits after pregnancy of about 9 months.

Size: head-body up to 3ft, tail 8in; weight 22-55lb; males about 1½ times size of females.

Color: dark brown to near black; long coat with lighter band along flanks.

Lifespan: up to 12 years.

►The wolverine covers large distances in the search for prey. Its sharp sense of smell finds carrion from far away.

The wolverine is the heaviest of the land-living members of the weasel family. It lives in the most northerly lands and is active all year, even in the depths of the Arctic winter.

Because of its warm coat, the wolverine is sometimes trapped for its fur. Over-hunting and human disturbance have reduced numbers in some areas, but these animals have always been thinly scattered.

WOLVERINE WANDERERS

In the course of a year a wolverine's movements cover a huge area. A male's territory can span up to 400sq miles, an area as big as some medium-sized counties. Female territories are one third the size or less and may overlap with those of males.

Wolverines keep others of the same sex away by scent marking, using urine, droppings or secretions of the scent glands. They may fight if they meet. Wolverines need to be well spaced to find enough food in their barren surroundings.

POWERFUL PREDATOR

The wolverine is enormously strong. It is also intelligent and determined. Sometimes it drives off larger animals from their prey, such as a solitary wolf from a sheep. It can kill caribou and other large deer. It drags down large prey by jumping on their backs and holding on. It can give a neck bite that kills a smaller mammal and disables a big one.

Large deer mostly fall prey to the wolverine in the winter, when it has the advantage in moving over snow. In summer it catches smaller mammals, eats nesting birds and their eggs and digs out wasps' nests for grubs. Some berries and other plants are also part of the diet, and at all times of the year the wolverine finds carrion.

The wolverine has an alternative name, glutton, which comes from its supposed greedy habits. In fact the wolverine will often bury or hide

▲ The wolverine is heavy and powerful, almost bear-like in appearance. It can climb trees fast and runs with a tireless, loping action.

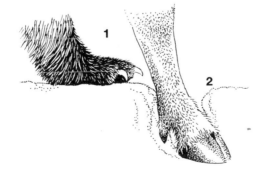

▼ The broad paws of the wolverine spread its weight and stop it sinking in snow (1). On soft snow the narrower feet of a deer sink in (2). So in winter the wolverine can catch deer. On hard ground in summer, deer can usually outrun the wolverine.

some of its food and come back to it in time of need. This may be as much as 6 months later. In the meantime the food is preserved in a natural deep freeze. A breeding female makes good use of such stores.

KIT CARE

Wolverines mate during the summer. The babies ("kits") are born at the end of the winter. They are blind and helpless. The mother gives birth in an underground den, often dug beneath a snowdrift.

The kits stay in or near the den until early summer, although the mother may move them to a new den if disturbed. The young emerge from the den with lighter coats than adults. They forage with the mother until the fall by which time they are nearly full grown and have an adult coat. Once they leave their mother, wolverines live solitary lives. The animals first breed when 2 years old.

SKUNKS

A female Striped skunk is out on a night's foraging. She has dug some beetles from the soft earth. Now she is at the edge of a wood, listening and waiting for the chance to catch a young rabbit. Another animal, a fox, steals along the edge of the wood. The skunk does not run away from it. She stamps in annoyance and walks stiff-legged in the moonlight. She is easy to see, holding her bushy tail straight up. The fox hesitates, then quickly goes past, keeping well clear of the skunk.

SKUNKS Mustelidae; sub-family Mephitinae (*13 species*)

■ Diet: insects, small mammals, eggs, some fruit.

◯ Breeding: 2-9 young after pregnancy of 42-66 days.

Size: head-tail 16-28in; weight 1-6½lb.

Color: mainly black, with white stripes or spots.

Lifespan: up to 10 years.

Species mentioned in text:
Hooded skunk (*Mephitis macroura*)
Pygmy spotted skunk (*Spilogale pygmaea*)
Striped skunk (*Mephitis mephitis*)

▲ Habitat: woods, open country, desert.

All species of skunk live in North, Central or South America. There are three main types. Seven species of hog-nosed skunk live in the southern USA, Central and South America. They have a long bare muzzle and large claws, which are both adaptations to digging.

The four spotted skunks live in the USA and Central America. They are light in build and are good climbers. Spotted skunks are striped, but these stripes may be broken into spots. The Striped skunk is a common species in North America and northern Mexico. The Hooded skunk is found in the south-west USA.

MAKING A STINK

The weasel family all have anal scent glands, but these are best developed in the skunks. Skunks are able to spray the contents of the two glands in the direction of an enemy. The spray is aimed at the face and causes intense irritation of the skin – sometimes even temporary blindness.

The spray is accurate over 6½ft or more. It contains sulfur compounds with an unbearable and very clinging smell. Although this is a good deterrent against most enemies, Great horned owls still attack skunks. Other animals may also attack if desperate with hunger. But usually a skunk is safe and not afraid of these predators. Skunks and their dens do not smell of the powerful spray.

THREAT DISPLAYS

Skunks do not hide from enemies. Instead they put on a warning display.

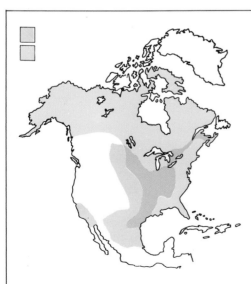

◄ Skunks are common carriers of the disease rabies in the USA. Infected skunks have the rabies virus in their saliva and tend to bite almost anything that moves. The map shows where rabies occurs in skunks and Red foxes in North America.

► Most animals avoid a skunk that gives a threat display or sprays. Rabid foxes, though, may still attack it, so infecting the skunk and spreading the disease.

Together with their obvious black and white color this is usually enough to put off an enemy. Only as a last resort does the skunk shoot its stinking spray.

The ability to spray scent develops at only a month old. When a skunk is born it is blind and hairless, but even then the skin shows signs of the pattern that will develop in the fur. The Striped skunk's eyes open at about 3 weeks, by which time the fur has grown. At 5 weeks the young begin to move around with the mother, finishing suckling at about 2 months. By the fall after their birth (usually in May) the young are looking after themselves.

▲The Pygmy spotted skunk is a rare species found only on the Pacific coast of Mexico. Spotted skunks have silky fur.

Skunks are found in many habitats, including towns. They are basically meat-eaters, but consume almost any food. They rest up in dens in bad weather. Skunks suffer from fleas, ticks and flatworms and from diseases such as rabies and distemper.

▼**Species of skunk A Western spotted skunk** (*Spilogale gracilis*) **(1)** does a handstand, a threat made before spraying its scent. The Hooded skunk **(2)** of the south-west USA. The Hog-nosed skunk (*Conepatus mesoleucus*) **(3)** has a long bare snout. In the very common Striped skunk**(4, 5)** the white stripes vary in number and thickness.

OTTERS

Off the Californian coast a Sea otter dives to the sea bed. It digs out a clam and a stone, then returns to the surface. The otter lies on its back at the surface balancing the stone on its chest. Taking the clam in one hand, it bangs it down repeatedly on the stone. The clam shell cracks, and the otter feeds on the flesh inside. Then it washes off the remains by rolling over. The stone stays in the otter's hands, to be used again.

As well as the Sea otter of the North Pacific there are another 11 species of otter found across most continents except Australasia and Antarctica. Several species venture into the sea, but most prefer the rivers, lakes and marshes. Otters can move across country, but like to be near water, where they find food.

HUNTING SESSIONS
Otters have tight-packed underfur with long guard hairs. The coat repels water and soon dries. An otter's body

is very supple. The tail is thick and muscular at the base, flattened below and in some species flattened above too. It helps in swimming. Most kinds of otter have webbed paws.

Otters are active and energetic. Many species have several hunting sessions in a day. An otter may eat daily 2lb of food spread over several meals. The teeth are very strong, helping otters crush the bones of their

prey. They digest their food quickly, giving them boundless energy.

MANY VOICES
Most otters, such as the European and North American river otters, live singly except when breeding. Then a

▼An Oriental short-clawed otter shows its streamlined shape under water, when its ears and nostrils are closed.

OTTERS Mustelidae; sub-family Lutrinae (*12 species*)

○ ■ 🦦

Habitat: aquatic (including sea) and on land.

Diet: fish, frogs, crabs, crayfish, shellfish.

○ Breeding: 1-5 (usually 2) young after pregnancy of 60-70 days (plus delay in some).

Size: smallest (Oriental short-clawed otter): head-body 16-24in, tail 12in, weight 11lb; largest (Giant otter): head-body 3-4ft, tail 2ft, weight up to 65lb.

Color: brown, some grayish.

Lifespan: up to 12 years.

Species mentioned in text:
Cape clawless otter (*Aonyx capensis*)
European river otter (*Lutra lutra*)
Giant otter (*Pteronura brasiliensis*)
North American river otter (*Lutra canadensis*)
Oriental short-clawed otter (*Aonyx cinerea*)
Sea otter (*Enhydra lutris*)

▲The Cape clawless otter uses its hands to catch and eat prey. Its long whiskers help it sense moving prey.

pair may stay together for just a few months. Other otters, such as the Cape clawless otter, live in pairs all the time. The most social otters include the Giant otter and the Oriental short-clawed otter. These move around in larger groups based on families.

Otters make many sounds. They chirp or bark to keep in touch and make a chattering sound when close. They also growl. Another way otters communicate is by scent. They may deposit droppings, urine or scent from the anal glands. These scents mark out territories and give information to other otters about the animal that left them. Groups of Giant otters in Brazil make communal scent areas on river banks, clearing all the plants in a wide semicircle.

▶Clever hands The Oriental short-clawed otter (1, forepaw a) catches prey with its hands and is good at feeling and grasping, as is the Cape clawless otter (b). The Spot-necked otter (*Hydrictis maculicollis*) (2) has webbed fingers (e) and catches food with its mouth. So does the Giant otter (c). The Indian smooth- coated otter (*Lutrogale perspicillata*) (3), has webbed fingers (d) but can hold a shell to its mouth. North American river otter (4,f) has a bare nose. Species can be told apart (i-viii) by the nose's shape.

BADGERS

A badger out hunting moves along a familiar path. At intervals he squats and leaves scent from the glands under his tail. It is a damp night, and he finds many worms to eat. Returning to his burrow he meets other animals of his group. They scent-mark each other.

Six species of badger, among them the ferret badgers, are found only in Asia. The Eurasian badger lives in Europe and Asia, the American badger just in North America. The ratel or Honey badger is widespread in Africa and across Asia as far as India.

Badgers have well developed anal glands. Ratels can produce a foul liquid to deter enemies, and the teledu or Malayan stink badger can squirt its scent at attackers.

LARGE BURROWS
Badgers have powerfully built wedge-shaped bodies with a small head and a

short thick neck. The long snout is used to find food. Their sense of smell is the most important one. Their front feet have long claws and are used in digging out food and making burrows.

The Eurasian badger makes burrows ("setts") with several entrances and chambers. Larger setts may be hundreds of yards long with many entrances. These are usually the work of several generations of badgers.

▼ Two young Eurasian badgers peer from the entrance to their burrow. They do not venture out until 2 months old.

BADGERS Mustelidae; sub-families Melinae and Mellivorinae (*9 species*)

● ■

◢ Habitat: wood and forest, some in mountains or grassland.

◪ Diet: small animals, fruit, roots.

○ Breeding: 1-5 young after pregnancy of 3½-12 months, including delay.

Size: smallest (Oriental ferret badger): head-body 13-17in, tail 6-9in, weight 4½lb; largest (Eurasian badger): head-body 2-3ft, tail 6-8in, weight 26½lb.

Color: black, white and gray, some species yellowish or brown.

Lifespan: up to 15 years.

Species mentioned in text:
American badger (*Taxidea taxus*)
Eurasian badger (*Meles meles*)
Oriental ferret badgers (*Melogale orientalis*)
Ratel or Honey badger (*Mellivora capensis*)
Teledu (*Mydaus javanensis*)

HUNTING PARTNERS

Most badgers eat all kinds of food items. The ratel is especially fond of bee grubs and honey. A small bird, the honeyguide, alerts and attracts a ratel by its cries, leading it to a bees' nest. The ratel, which has enormous strength and a tough skin, rips open the nest, and both ratel and bird have a feast. American badgers are also said to form hunting partnerships with coyotes, and share a kill with them.

Some types of badger are solitary, but the Eurasian badger lives in groups of up to 12 related animals. These relatives share a main burrow and outlying burrows within their territory. They use special dungpits to deposit scent, droppings and urine. Pits at the edge of the territory help keep out intruders.

A female Eurasian badger makes a nursery chamber in the burrow for the birth of the cubs at the end of winter. At birth the cubs are pale, but are striped before leaving the den.

▲The American badger lives in open country, eating mainly meat. This one has caught a rattlesnake.

▼The teledu (1) lives in Sumatra, Borneo and other nearby islands. It is a ground dweller with a tiny tail. The Oriental ferret badger (2) has a long tail and sometimes climbs trees. It lives in forests in Java and Borneo.

CIVETS

At night, in the humid rain forests of West Africa, there is a constant babble of noise. But every now and then, this wall of noise is interrupted by a loud, ghostly cry, like the hooting of an owl. Similar cries reply from half a mile away. These are African palm civets calling to one another.

Civets and their relatives, the genets and linsangs, are cat-like mammals. Most live in the hot, moist areas of Africa or Asia, though one kind, the Common genet, lives in southern France, Spain and Portugal. They are animals of the night, so have large eyes. They also have a good sense of smell and excellent hearing. Long tails enable them to keep their balance when climbing trees. They live mainly in forests, and eat insects, small mammals and fruit. The Bear cat or binturong of South-east Asia has a powerful, long tail which it uses to hang on to branches.

There are eight species of palm civet, one in Africa and seven in Asia. They all spend most of their time in trees, where they are skilled climbers, getting a grip with strong, curved claws. Palm civets eat many kinds of fruit, even some which are poisonous to humans. The Common palm civet of Asia eats the fruit of at least 35 types of plant. Like all the other palm civets, it also eats snails, scorpions, birds, rats and mice. This species is very common, often living close to people, and in some areas it is a pest.

On the other hand, the Otter civet and Jerdon's palm civet, both from Asia, and the Large spotted civet, from South Africa, are now all very rare and endangered species.

A WORLD OF SMELLS
A male palm civet has a special home area or territory which he patrols regularly. He marks the edge of his area with a scent made in a fold of skin at his rear end. People collect the scent from captive animals for use in expensive perfumes. A strong, healthy male may have one to three females in a large home area. He visits the females' own areas every few days.

THE PERFECT KILLER
Genets feed on insects, lizards, small birds, rats and mice. They have sharp teeth and excellent sight. They live mainly in trees, but also hunt on the ground. Genets are such good hunters that in Europe the Common genet was once kept as a rat-catcher.

▲A Giant or Celebes palm civet eating a young shoot. This rarely seen civet lives only on the island of Sulawesi (Celebes).

►Types of civet An African linsang (*Poiana richarsoni*) (1) eats a nestling bird. The Banded palm civet (*Hemigalus derbyanua*) (2) devours a lizard, while an Oriental civet (*Viverra tangaluna*) (3) smells an enemy and raises its crest of bristly hairs. An ever-curious Common palm civet (4) sniffs the air for signs of a mate. Using its tail for support, a Bear cat or binturong (5) searches for fruit.

CIVETS Viverridae (*35 species*)

● ■ ☠

Habitat: mainly in trees in forests, scrub, mountains; also along river banks.

Diet: varied, including fruit, leaves, shoots, insects, small lizards, frogs, birds, rats, mice.

Breeding: litter of 1-3 after pregnancy of 70-90 days.

Size: smallest (African linsang): head-tail 28in, weight 1½lb; largest (African civet): head-tail 50in, weight 28½lb.

Color: dark spots and/or banding on a pale, sandy background.

Lifespan: 15-34 years.

Species mentioned in text:
African civet (*Civettictis civetta*)
African palm civet (*Nandinia binotata*)
Bear cat or binturong (*Arctictis binturong*)
Common genet (*Genetta genetta*)
Common palm civet (*Paradoxurus hermaphroditus*)
Giant or Celebes palm civet (*Macrogalidia musschebroekii*)
Jerdon's palm civet (*Paradoxurus jerdoni*)
Large spotted civet (*Viverra megaspila*)
Otter civet (*Cynogale bennettii*)

MONGOOSES

At dawn a pack of 14 Banded mongooses leave their termite-mound den. They move off in single file, then fan out to search for beetles. They keep in contact by calling to one another. An adult has stayed in the den to guard the group's 10 babies. Hours later the pack returns. Mothers suckle babies, while several younger adults bring them beetles. Then the main pack sets off after food again.

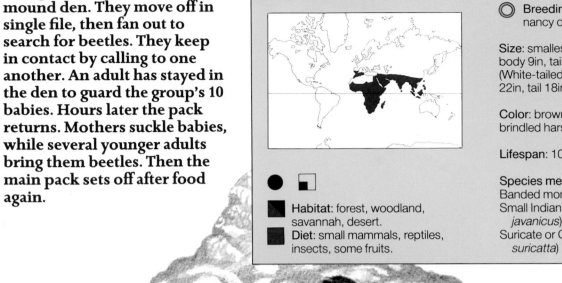

MONGOOSES Viverridae; sub-families Herpestinae, Galidiinae (*31 species*)

○ **Breeding:** 2-4 young after pregnancy of 42-105 days.

Size: smallest (Dwarf mongoose): head-body 9in, tail 7in, weight ½lb; largest (White-tailed mongoose): head-body 22in, tail 18in, weight 11lb.

Color: brown to yellowish, grizzled or brindled harsh fur.

Lifespan: 10 years.

Species mentioned in text:
Banded mongoose (*Mungos mungo*)
Small Indian mongoose (*Herpestes javanicus*)
Suricate or Gray meercat (*Suricata suricatta*)

● ◻

■ **Habitat:** forest, woodland, savannah, desert.
■ **Diet:** small mammals, reptiles, insects, some fruits.

Twenty species of mongoose are African. Four more live in Madagascar. Seven species live in South Asia. They are often the commonest carnivores in the places they live. Agile and active, some climb, but most species live on the ground.

SNAKE KILLERS?

Mongooses are famous for killing snakes and some really do kill venomous snakes. Mongooses are fast and alert, and tire less quickly than a reptile. They dodge the snake's fangs until they can jump in with a killing bite to the neck.

But most mongooses feed on easier prey, taking a whole range of small

◄Suricates scan their surroundings in the Kalahari desert. These mongooses live in groups and are active by day.

▼**Mongoose species** Adult Dwarf mongoose (*Helogale parvula*) **(1)** feeding a youngster. Selous' mongoose (*Paracynictis selousi*) **(2)**. Narrow-striped mongoose (*Mungotictis decemlineata*) **(3)**. Egyptian mongoose (*Herpestes ichneumon*) **(4)** about to break an egg on a rock. Marsh mongoose (*Atilax pauludinosus*) **(5)** scent-marks a stone.

White-tailed mongoose (*Ichneumia albicauda*) **(6)**. Ring-tailed mongoose (*Galidia elegans*) **(7)**.

▶Banded mongooses work together as a pack when feeding. As a group they can repel large enemies like jackals.

animals. Some kinds are especially fond of insects. Others eat crabs and frogs. Many mongooses can crack open eggs by throwing them against stones before lapping up the very nutritious contents.

MONGOOSE GROUPS

Most mongooses live by themselves or sometimes in pairs. Most hunt stealthily by night. A few species such as the Banded mongoose live in close-knit groups and are active by day. Group members defend each other from attacks by predators.

The Small Indian mongoose was introduced to Hawaii and the West Indies in the hope it would kill rodent pests. Unfortunately it killed many native animals too and is now thought of as a pest itself.

HYENAS

A snapping, snarling, cackling pack of Spotted hyenas is a nightmarish sight. These animals lope through the darkness in search of young or weakened animals. They use their strong jaws to pull their prey to the ground, where it is killed within seconds. Hyenas are not just humble scavengers but efficient killers.

Hyenas are dog-like mammals living in Africa and parts of Asia. They have a short, shaggy mane and a bushy tail. The tail is held upright when the animal is hunting, but down between the legs when it is running away from an enemy such as a lion. Male and female hyenas look the same, though the females are slightly larger. Hyenas have sloping backs, which gives them a lop-sided appearance. The neck and shoulder muscles are very strong, enabling hyenas to pull much larger animals to the ground.

Hyenas hunt mainly in the dark, so they have very good night vision. They can run at speeds of up to 36mph for several miles. A hyena can catch and eat a small animal in a matter of seconds. In order to catch larger and

▼Hyenas communicate by scent-marking from the anal pouch, as in this Brown hyena (1), or, as in these Striped hyenas, by raising their shoulder crests in greeting (2) or sniffing each other (3).

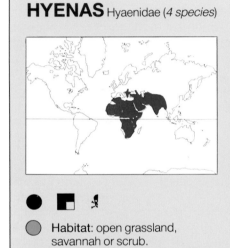

HYENAS Hyaenidae (*4 species*)

Diet: animal flesh, grubs, fruit, eggs.

Breeding: litters of 2-4 after pregnancy of 60-110 days.

Size: head-tail 3-5½ft; weight 18-175lb.

Color: yellow to brown or red, with black spots and stripes.

Lifespan: 13-25 years.

Species mentioned in text:
Aardwolf (*Proteles cristatus*)
Brown hyena (*Hyaena brunnea*)
Spotted hyena (*Crocuta crocuta*)
Striped hyena (*Hyaena hyaena*)

Habitat: open grassland, savannah or scrub.

1

2

3

◄A pack of Spotted hyenas crowds round a carcass. One animal may eat 33lb of flesh in one meal.

▲A frustrated Spotted hyena tries to break open an ostrich egg.

more powerful animals, hyenas hunt in packs of between 10 and 20. They can eat a zebra carcass in 15 minutes. If there is too much food, uneaten parts of the animal may be dragged off and hidden in water-holes.

Spotted hyenas have very large cheek teeth for crushing bone and tearing through tough hide. They eat all parts of an animal except for the hoofs and hair. Undigested bone passes out as a white powder. Striped and Brown hyenas have smaller cheek teeth. They hunt smaller animals, but catch less of their own food, relying more on scavenging dead animals and the remains of other animals' meals. Sometimes a hyena pack will drive away a lion from its kill. Hyenas also eat birds' eggs, insects, fruit and vegetables.

A type of hyena known as the aardwolf feeds almost entirely on termites, which it detects by sound and scent. This shy and rarely-seen creature has a slender muzzle and pointed teeth. It licks up its termite prey with a long sticky tongue. The aardwolf is a solitary forager. It does not need the help of others to find food.

▶The types of hyena
A Striped hyena scavenging (1). Young Brown hyenas playing (2). An aardwolf hunting termites (3). Spotted hyenas hunting a zebra (4).

LIVING TOGETHER

Brown hyenas live in groups of 4 to 14 in areas where there is little food, so they need large territories in which to hunt. Spotted hyenas live where there is much more prey so they join up in groups of 30 to 80 in smaller areas. Hyenas communicate by calls and scents. The Spotted hyena is often called the "laughing hyena" because of the high-pitched cackle it makes.

Hyenas have a special pouch at the base of the tail which produces a black and white scented paste. They spread this on to grass stems to mark their home range. The paste from each individual has a unique smell. When animals of the same kind meet, they can tell whether they belong to the same group or not by sniffing each other's pouches.

RAISING YOUNG

Female hyenas raise their young without help from the males. Spotted hyenas usually give birth to twins. The newborn are well-developed and covered with hair. They live with their mother in an underground den shared by several females, where they are suckled for up to 18 months. Then the young leave the den to fend for themselves.

1

2

3

4

MARSUPIAL CARNIVORES

The Moon shines down brightly on a group of sand-hills in the Australian outback. From a shallow burrow comes a buff colored "mouse" about 6in long, with a 4in black-tipped tail. It is a mulgara. Peering into crevices, it searches busily. In one hole it finds a beetle, which it flicks out with a front paw and snaps up. Finding a lizard nearly as large itself, it attacks it savagely, biting the lizard behind the head. After feeding, for a while its hunger is satisfied. Nearby, still searching for her food, is a female mulgara, dragging six little pink young attached to her teats. Morning comes. Before it is hot, the mulgaras are back home underground.

The marsupial carnivores are a large family found in many habitats through Australia and New Guinea. They live in tropical rain forest and light woodlands, up mountains and even by the coast. Many kinds are able to survive in deserts.

Many of the species look rather similar. Small species are often mouse-like in appearance, although their habits are very different to true mice. Larger species are the equivalents to the stoats, civets and cats of other parts of the world. Instead of the mouse-like coloration of the small carnivores they are red-brown or black and may have spots of white.

SAVAGE ATTACK
Over half of the marsupial carnivores are animals weighing less than 3½ ounces. Some are less than a tenth of this and are among the smallest mammals. But they have sharp teeth and attack their prey savagely. The Common planigale, little more than

MARSUPIAL CARNIVORES
Dasyuridae (*51 species*)

Size: smallest (Pilbara ningaui): head-body 2in, tail 2½in, weight ¹⁄₁₂ ounce; largest (Tasmanian devil):head-body 26in, tail 10in, weight 17½lb.

Color: many smaller species grayish, mouse-colored; larger species often brown or black with white markings.

Lifespan: 11.5 months (Brown antechinus) to 7 years (Tasmanian devil).

Species mentioned in text:
Brown antechinus (*Antechinus stuartii*)
Common planigale (*Planigale maculata*)
Dibbler (*Parantechinus apicalis*)
Dusky antechinus (*Antechinus swainsonii*)
Eastern quoll (*Dasyurus viverrinus*)
Fat-tailed dunnart (*Sminthopsis crassicaudata*)
Mulgara (*Dasycercus cristicauda*)
Pilbara ningaui (*Ningaui timealeyi*)
Spotted-tailed quoll (*Dasyurus maculatus*)
Tasmanian devil (*Sarcophilus harrisii*)
Tasmanian wolf (*Thylacinus cynocephalus*)

● ▢ 🦴
◨ Habitat: forest, stony desert, mountain heath.

◼ Diet: insects, worms, spiders, lizards, mice, other small animals, carrion; some also eat flowers and fruit.

○ Breeding: 2-12 young after pregnancy of 12-27 days.

▼Four month old Eastern quolls are too old to stay on their mother's teats. She leaves them in a grass-lined nest. Six tiny young fix to the teats in the pouch at birth and stay there for 10 weeks.

►The Dusky antechinus usually lives on the ground. Like many of the small "mouse-like" marsupials it is a fierce predator.It catches insects and small vertebrates such as lizards.

2in long, plus a 2in tail, can overcome grasshoppers as big as itself.

Many of the marsupial carnivores feed mainly on insects. Even those that are cat-sized rely to a large extent on beetle larvae, although they can catch and kill reptiles, birds and mammals. The mulgara kills its prey by biting and shaking. The final bite is always at the head or neck.

Most marsupial carnivores hunt by stealth or surprising prey. Few chase prey far. One exception is the Spotted-tailed quoll of Tasmania, a large cat-like animal with a body up to 30in long and a long tail. It is active and agile and can run down prey. It takes wallabies, gliding possums and reptiles. It also kills with an accurate and powerful bite to the back of the head.

Spotted-tailed quolls are sometimes

◄A Tasmanian devil shows its strong jaws and sharp teeth. It can chew and swallow all of a carcass, even bones.

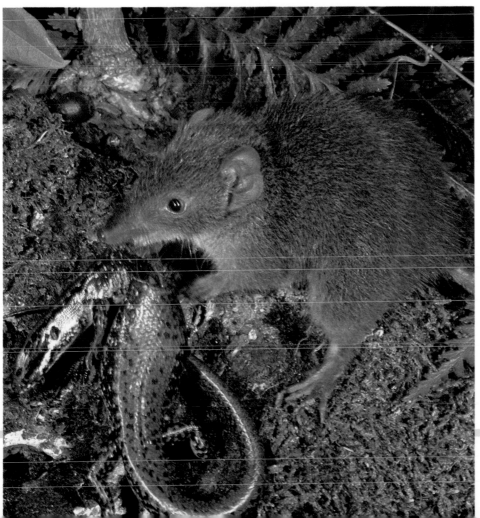

▲The Tasmanian wolf or thylacine was one of the largest marsupial carnivores. Its fossil bones are found in Australia, but in historical times it lived only in Tasmania. It looked like a dog, about 24in high at the shoulders, but fell away to the hind legs. It had a long stiff tail and a coat that was sandy with darker stripes. It had teeth like a dog and caught animals such as wallabies. It was said to kill sheep, so people hunted it. The last known specimen was caught in 1933 and died in 1936.

known as tiger-cats, because of their size and behavior. They make sudden hisses when alarmed, and an angry animal produces unnerving screams. Quolls are mainly active at night. They are able to track prey by following scent trails. These animals are relatively long-lived, becoming adult at 2 years old, and living 6 years or more.

FEEDING ON CARRION

The Tasmanian devil, the largest living marsupial carnivore, is black, with a white mark across the chest. It has a rather tubby body and short legs. It is sometimes said to look like a small bear. In habits it is really more like a hyena. Although rarely able to kill active prey, it is very good at eating carrion. It can crunch up dead animals' bodies, bolting large lumps. It has been seen cramming intestines into its mouth with its front paws as though it was eating spaghetti.

DESERT AND TREE SPECIES

Many small carnivores are able to survive well in the Australian deserts. The mulgara, for example, can exist where there is no drinking water. All water must come from the insects and small vertebrates it eats. It holds on to water by producing very concentrated urine. It cuts down on evaporation of water from its body by living underground during the day. This avoids the high daytime temperatures.

In dry areas where food is only plentiful for part of the year carnivores may store food. The Fat-tailed dunnart has a tail which it uses for storing fat. It uses the fat up in time of need. When food is scarce it may also save energy by becoming inactive and allowing its body to slow right down. Even the body temperature drops.

Many marsupial meat-eaters, such as the quolls or "native cats," can climb well when the need arises, although they may spend much time on the ground. The phascogales are smaller

species that are good climbers. They are usually in trees, either in forests or in open woodland.

Phascogales have long tails, the end-part of which is a brush of long hairs. The tail is used for balance and also probably as a signal. Phascogales are as agile as squirrels and are mainly active at night.

MATING AND BREEDING

Some of the small carnivores have a very short lifespan. Females of the Brown antechinus and related species may live for over a year, but the males are "annuals." They are born in early spring and are weaned by mid-summer. In late winter comes the mating season. The antechinus males are frantically active, searching for females and mating with them. Then they all die. At the end of the winter only pregnant females are left to carry on the species.

TOO MANY BABIES

Like kangaroos, the marsupial meat-eaters have their babies born in a very undeveloped stage after a short pregnancy. But few have a well developed pouch to hold the babies. In most cases there is just a fold of skin.

The exposed young are carried attached to the mother's teats for the first weeks of life.

There may be more babies born than the mother has teats. Those that do not find a teat to attach to will die. Some species, for example the planigales, have as many as 12 teats and may produce 20 babies.

SHRINKING NUMBERS

Since Europeans settled in Australia many carnivores have decreased in both distribution and numbers, often because habitat has been destroyed or used for farming. The larger species, such as the quolls, have suffered most. The Eastern quoll, for example, is endangered in parts of its range, although this may be due to competition from domestic cats that have become wild. Five species are believed in danger of extinction. These include the dibbler, a little mouse-like species which may already be extinct.

▼**A range of marsupial carnivores**
Kultarr (*Antechinomys laniger*) **(1)**.
Pilbara ningaui **(2)**. Three-striped
marsupial mouse (*Myoictis melas*) **(3)**.
New Guinea marsupial cat (*Satanellus
albopunctatus*) **(4)**. Fat-tailed or
Red-eared antechinus (*Pseudantechinus
macdonnellensis*) **(5)**. Marsupial mouse
(*Phascolosorex dorsalis*) **(6)**. Red-tailed
phascogale (*Phascogale calura*) **(7)**.
Little red antechinus (*Dasykatula
rosamondae*) **(8)**. Long-tailed marsupial
mouse (*Murexia longicaudata*) **(9)**.
Fat-tailed dunnart **(10)**. Common
planigale **(11)**.

69

DOLPHINS

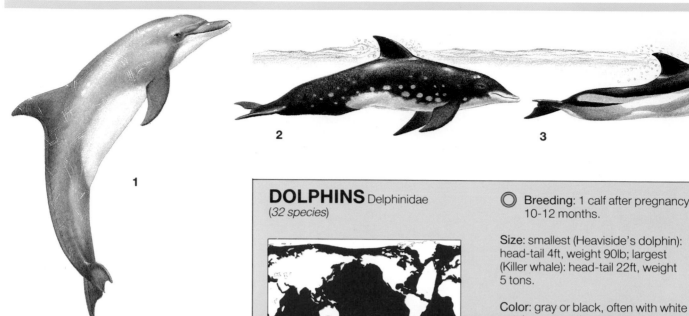

1

2

3

A group of dolphins have found a shoal of fish. They spread out round the edge of the shoal, leaping out of the water and diving back. They herd the fish into a tighter bunch. Now the dolphins feed well. They snap up fish one after the other. At last they are full. They begin to play, chasing, leaping and spinning.

DOLPHINS Delphinidae
(*32 species*)

○ ■

▨ Habitat: mostly coastal shallows, some open ocean.

■ Diet: fish, squid, other animals.

○ **Breeding**: 1 calf after pregnancy of 10-12 months.

Size: smallest (Heaviside's dolphin): head-tail 4ft, weight 90lb; largest (Killer whale): head-tail 22ft, weight 5 tons.

Color: gray or black, often with white patches, sometimes with other patches of color.

Lifespan: 20-50 years.

Species mentioned in text:
Bottle-nosed dolphin (*Tursiops truncatus*)
Dusky dolphin (*Lagenorhynchus obscurus*)
Heaviside's dolphin (*Cephalorhynchus heavisidii*)
Killer whale (*Orcinus orca*)

▲▶**Species of dolphin in common poses** Bottle-nosed dolphin (**1**). Rough-toothed dolphin (*Steno bredanensis*) (**2**). Atlantic white-sided dolphin (*Lagenorhynchus acutus*) (**3**). Spotted dolphin (*Stenella plagiodon*) (**4**). Common dolphin (*Delphinus delphis*) (**5**). Northern right whale dolphin (*Lissodelphis* *borealis*) (**6**). Dusky dolphin (**7**). Atlantic humpbacked dolphin (*Sousa teuszii*) (**8**). Melon-headed whale (*Peponocephala electra*) (**9**). Commerson's dolphin (*Cephalorhynchus commersoni*) (**10**). False killer whale (*Pseudorca crassidens*) (**11**). Killer whale (**12**). Risso's dolphin (*Grampus griseus*) (**13**).

13

Dolphins are small whales. They are found in all the world's oceans. In most dolphins the jaws form a well-developed beak. Above this there is usually a "melon," a protruding rounded forehead. The nose is not on the beak. A dolphin breathes through a blowhole up on top of the head above the melon. Dolphins have a single dorsal fin which curves backwards.

Dolphins belong to the side of the whale family tree known as toothed whales. The description fits, because most have between 100 and 200 teeth in their jaws. Some have as many as 224. The teeth are all similar and sharply pointed. They are ideal for holding slippery prey.

Most dolphins feed on fish. Some prefer squid, and others will even eat shrimps. Many kinds of dolphin make use of shoals of fish swimming near the surface, but others will also dive deep for prey, or even pick fish from the sea bottom.

SOUND SENSE
Dolphins have good hearing and make many sounds themselves. Hearing is very important to them, both for keeping in touch with one another and for catching prey. Dolphins make some clicking sounds and whistles we can hear, and also other sounds which are much too high for the human ear. Some of the whistles are made when they are in particular moods or doing particular things. These can give information to other dolphins.

The very high sounds are used to beam out in front of the dolphin and produce echoes from objects. The dolphin hears the echoes and from them can tell what is around. Dolphins especially use this system to find prey. The melon on a dolphin's forehead helps to focus the sound. Although we cannot hear them, some high-pitched sounds produced by dolphins are very loud. They may frighten and confuse prey, and perhaps even stun them.

MERCILESS HUNTER
The biggest of all the dolphins is the Killer whale. It is long-lived and intelligent. It is also one of the fastest swimmers, able to travel at 25mph. The Killer whale is widespread but is commonest in cool seas where there is plenty of prey. It lives in groups called pods of up to 40 individuals which know one another and are able to co-operate in hunting.

The Killer whale eats squid and fish, including sharks. It also kills seals, walruses and porpoises and may even attack larger whales. One Killer whale is recorded as having the remains of 15 seals and 13 porpoises inside it. Killer whales have been seen tipping ice-floes to catch seals as they fell off. Although it is so strong, there is no

71

▲ Dolphins can be inquisitive and playful. Here Bottle-nosed dolphins investigate two odd creatures at the edge of the sea.

record of a Killer whale making an unprovoked attack on a human.

GATHERING IN GROUPS

Killer whales stay together in their groups for life. In a group there is likely to be an adult male, three or four adult females and some younger whales of both sexes.

Other dolphins live in groups too, but they are often less fixed than those of Killer whales. The pair and calf, or mother and calf, keep together, but may join or leave bigger groups. Species that live inshore may form herds of 12 or so, and sometimes larger numbers come together where feeding is good. Some ocean dolphins form herds of 1,000 or even 2,000 at feeding areas. Dolphins are able to co-operate in hunting, driving the shoals of fish together.

Some dolphins live in small individual areas, such as the Bottle-nosed dolphin, which may keep within about 35sq miles. The Dusky dolphin is very different – it may roam over 600sq miles.

BORN BACKWARDS

Mating and birth can take place at any time of year, but in many species most births take place in summer. A baby dolphin is born underwater, tail first. As soon as the head is out, the baby must be got to the surface to take a breath and fill its lungs. The mother, and often other female dolphins, help the baby do this.

Once it is breathing, the youngster can swim, but the mother and "aunts" are very protective. Dolphin babies suckle underwater between breaths. The mother's milk is very fatty. She pumps it quickly into the baby.

BIG BRAINS

Dolphins have large brains compared to their body size. In animals this is usually a sign of intelligence. Dolphins can learn tricks readily, can remember complicated routines and can mimic some sounds and actions. It is doubtful, though, whether they are really much more intelligent than some other mammals such as dogs or elephants. Much of the large brain seems to be for dealing with the sounds that are so important to a dolphin.

DOLPHINS AND PEOPLE

Dolphins are curious, and sometimes are interested in humans. There have been several instances of "friendships" being struck up between wild dolphins and people.

People, though, are not always good for dolphins. Many dolphins have been killed by fishing boats using large nets. The animals get tangled and drown. Each year in the 1960s and 1970s about 110,000 dolphins were killed this way in the Eastern Pacific alone.

Now tuna fishermen can use special nets which reduce the threat to dolphins. The nets have a panel of fine mesh furthest from the boat. Fleeing dolphins do not get tangled in this and can escape over the net rim. Some countries, such as the United States, use human divers stationed in the nets to help trapped dolphins.

▶ Killer whales have large dorsal fins. As well as eating fish, they feed on warm-blooded prey, including other dolphins.

PORPOISES

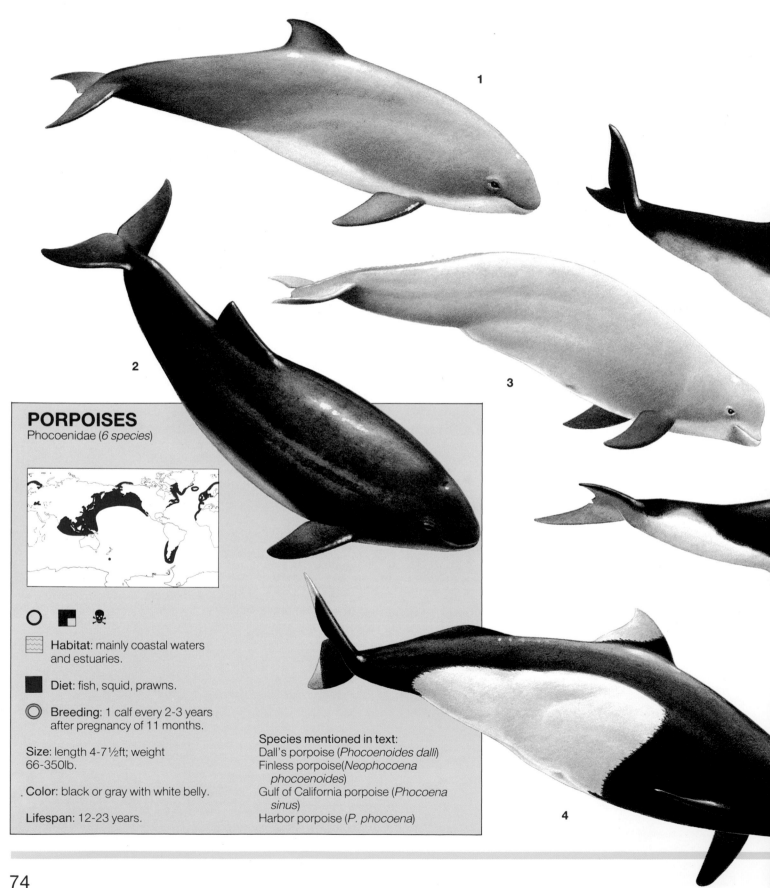

PORPOISES
Phocoenidae (*6 species*)

○ ◼ ☠

▨ **Habitat:** mainly coastal waters and estuaries.

◼ **Diet:** fish, squid, prawns.

◎ **Breeding:** 1 calf every 2-3 years after pregnancy of 11 months.

Size: length 4-7½ft; weight 66-350lb.

Color: black or gray with white belly.

Lifespan: 12-23 years.

Species mentioned in text:
Dall's porpoise (*Phocoenoides dalli*)
Finless porpoise(*Neophocoena phocoenoides*)
Gulf of California porpoise (*Phocoena sinus*)
Harbor porpoise (*P. phocoena*)

▲A Dall's porpoise ploughs through the sea. This species is the one most often attracted to boats.

6

5

◀The six types of porpoise The Gulf of California porpoise (1), endangered by competition with fishermen. The little-known Burmeister's porpoise (*Phocoena spinipinnmis*) (2). The Finless porpoise (3) lacks a triangular back fin. The strikingly-colored Dall's porpoise (4). The black eye-rings give the Spectacled porpoise (*Phocoena dioptrica*) (5) its name. The Harbor porpoise (6) is the most frequently seen species, but its numbers are in decline.

Drifting out on the north Pacific Ocean, some fishermen on a small boat are startled by the sudden appearance of a Dall's porpoise, which swims boisterously around them. They glimpse at the blunt "smiling" face, and hear a loud snorting. Then, with a flick of its tail, the porpoise speeds away through the waves.

Porpoises are streamlined, fish-shaped mammals related to whales and dolphins. Unlike dolphins, they do not have a long snout, but, like all whales, they breathe through a blow-hole behind the blunt head. They swim at great speed, using the flat tail for power and flippers for steering.

Porpoises usually live alone, mainly in warmer coastal waters and estuaries. During the breeding season, though, porpoises form small groups called schools. After mating, the male and female pairs split up and each female rears her young (calves) without the help of the male. The common Harbor porpoise begins breeding at the age of 5 or 6 years. Dall's porpoise matures later, at around 7 years. A female Harbor porpoise suckles her calf for about 8 months, while Dall's porpoise may produce milk for up to 2 years. The calf may stay with its mother for a year or two after weaning, until the mother becomes pregnant again. Young Finless porpoises often hitch a ride on their mothers' backs by holding on to a series of small ridges.

Females and calves may form small groups of four to six, sometimes with additional young males. The young males may eventually form their own small, all-male groups.

USING SOUNDS
Porpoises communicate with each other using a wide range of sounds, including clicks, squeaks and grunts.

They hunt fish using keen eyesight and echolocation. They make high-pitched sounds which bounce back off squid or small shoals of fish and which the porpoises hear. An adult Harbor porpoise needs 6½ to 11lb of food per day, while the larger Dall's porpoise eats 22 to 26lb daily. The Finless porpoise probably finds much of its food by digging about with its snout in the sandy or muddy bottoms of estuaries.

Harbor porpoises eat mostly herring, sardine and mackerel, while Gulf of California porpoises eat fish called grunts or croakers. The other species of porpoise probably live on mullet, anchovies and squid. All porpoises usually swallow their prey whole.

PORPOISE PROTECTION
Although porpoises are often seen in coastal waters and estuaries, little is actually known about their way of life. This makes their steady fall in numbers particularly worrying. We do not know how best to protect them. Porpoises, like whales, often "beach" themselves on shallow coastlines. It is possible then to examine the stomach contents of animals which die in this way to see precisely what they eat.

Fishermen also depend on the prey caught by porpoises, and many feel themselves to be in competition with these animals. Many porpoises become trapped in fishing nets because they cannot detect the fine mesh with their echolocation. They just follow the shoal of fish into the trap. They can be released unharmed if they are handled with care, but many are deliberately killed.

Some scientists would like to see all nets "labelled" with a device which makes a sound to warn off porpoises, but this would be very expensive. Chemical pollution of the seas also kills porpoises, and here the solution is simple and obvious. Only the will to do something about it seems lacking.

SPERM WHALES

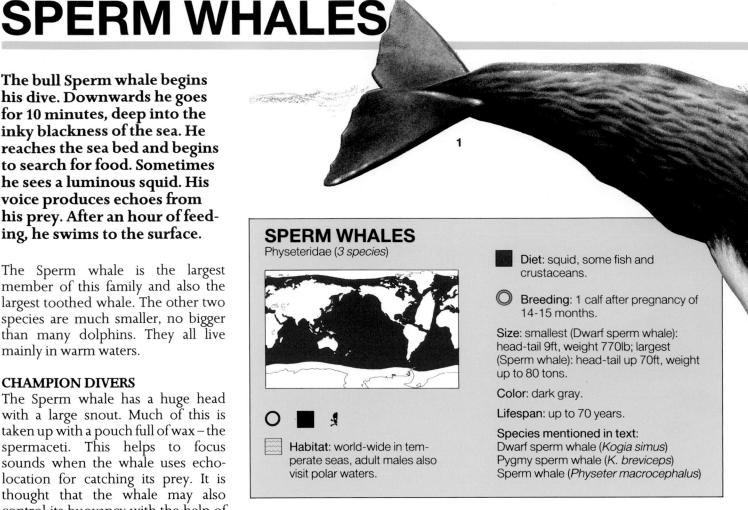

1

The bull Sperm whale begins his dive. Downwards he goes for 10 minutes, deep into the inky blackness of the sea. He reaches the sea bed and begins to search for food. Sometimes he sees a luminous squid. His voice produces echoes from his prey. After an hour of feeding, he swims to the surface.

The Sperm whale is the largest member of this family and also the largest toothed whale. The other two species are much smaller, no bigger than many dolphins. They all live mainly in warm waters.

CHAMPION DIVERS
The Sperm whale has a huge head with a large snout. Much of this is taken up with a pouch full of wax – the spermaceti. This helps to focus sounds when the whale uses echo-location for catching its prey. It is thought that the whale may also control its buoyancy with the help of the wax, cooling it to a solid for sinking, and warming it to liquid for rising.

Sperm whales make the deepest dives of all mammals. They have been picked up on sonar at 4,000ft depth. They have been found to have eaten bottom-living sharks in an area where the sea bed was 10,500ft down. Females can dive for an hour. The bulls, which are on average 13ft longer and twice as heavy, can dive for longer.

HUNGRY FOR SQUID
Sperm whales eat many things, but much of their food is squid. Most of their prey are about 3ft long, although some are smaller. One Sperm whale was found with 28,000 small squid inside it. Sperm whales sometimes also catch Giant squid, which are over 30ft long, although then they may be scarred by bites or sucker marks.

SPERM WHALES
Physeteridae (*3 species*)

○ ■ ☠

▨ **Habitat:** world-wide in temperate seas, adult males also visit polar waters.

■ **Diet:** squid, some fish and crustaceans.

◎ **Breeding:** 1 calf after pregnancy of 14-15 months.

Size: smallest (Dwarf sperm whale): head-tail 9ft, weight 770lb; largest (Sperm whale): head-tail up 70ft, weight up to 80 tons.

Color: dark gray.

Lifespan: up to 70 years.

Species mentioned in text:
Dwarf sperm whale (*Kogia simus*)
Pygmy sperm whale (*K. breviceps*)
Sperm whale (*Physeter macrocephalus*)

▼A pod (small herd) of Sperm whales rise to the surface to breathe. The blowholes are on the left of the heads.

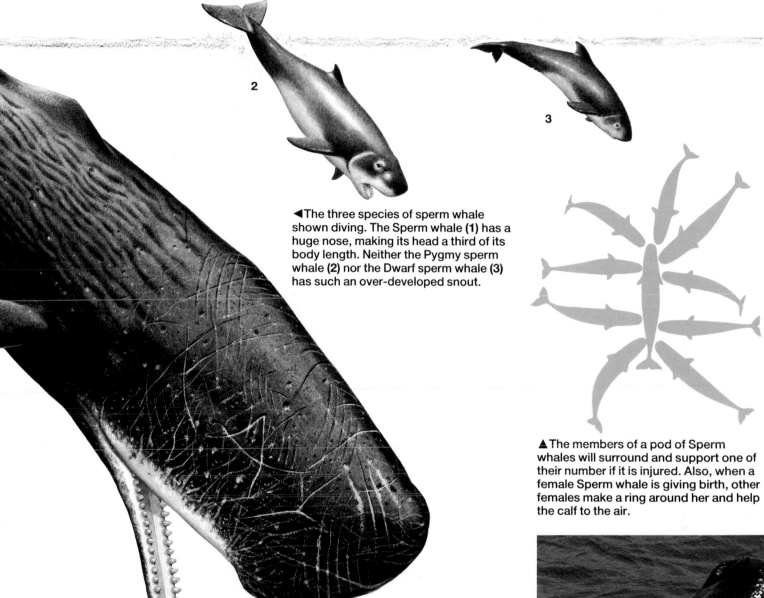

The three species of sperm whale shown diving. The Sperm whale (1) has a huge nose, making its head a third of its body length. Neither the Pygmy sperm whale (2) nor the Dwarf sperm whale (3) has such an over-developed snout.

▲The members of a pod of Sperm whales will surround and support one of their number if it is injured. Also, when a female Sperm whale is giving birth, other females make a ring around her and help the calf to the air.

TRAVELLING THE OCEANS

Sperm whales are commonest where ocean currents meet or water rises from the deep. Here there is plenty of food. The female Sperm whale lives in groups, and so do young males. Big bulls live alone except during the breeding season. Then they may fight for a harem.

Mating and birth take place near the equator. Afterwards the Sperm whale herds move to cooler water. The big males go much farther than others, as much as 4,900 miles to the cold waters near the North and South poles. Then all move back for the next breeding season.

About one-third of the weight of a Sperm whale is blubber that helps it keep warm in water. Blubber is a unique source of some lubricating oils, as is spermaceti. For these products, and for its meat to eat, the animal has been hunted relentlessly.

▶A Sperm whale calf breaks the surface, showing the dark wrinkled skin of this species.

WHITE WHALES

From high above, the scene looks peaceful as a herd of white whales go gliding through the clear water. But underwater there is a barrage of sound. The whales are belugas, and as they swim they call to one another. Squeals, clangs and whistles echo through the water. Sometimes there are chirps and mooing sounds. Their noises long ago earned them the sailors' name "sea canaries."

There are two species of white whale, the beluga and the narwhal. The beluga is always white as an adult. The narwhal has an unusual coloration, with a back covered in little patches of gray-green, black and cream, but this animal too whitens with age, from the belly upwards.

FLEXIBLE NECK
The beluga has a far more flexible neck than most whales and can turn it sideways to almost a right angle. It swims slowly on the surface, staying in shallow water. It dives to the bottom to find shoaling fish and crustaceans.

Groups of beluga sometimes work together to herd fish into shallow water or to a sloping beach, where they are more easily caught. A beluga can also hunt individual prey on the bottom. It can purse its lips and suck or blow water to move sand or dislodge prey. The beluga has up to 40 teeth, but they are not important in feeding. They may serve in visual threat displays and jaw-clap noise-making, used as communication.

▼A group of male narwhal show their iridescent gray-green-and-black backs as they swim in a dark blue icy sea.

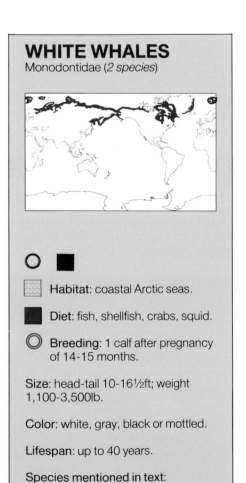

WHITE WHALES
Monodontidae (*2 species*)

○ ■

▨ Habitat: coastal Arctic seas.

■ Diet: fish, shellfish, crabs, squid.

◎ Breeding: 1 calf after pregnancy of 14-15 months.

Size: head-tail 10-16½ft; weight 1,100-3,500lb.

Color: white, gray, black or mottled.

Lifespan: up to 40 years.

Species mentioned in text:
Beluga (*Delphinapterus leucas*)
Narwhal (*Monodon monoceros*)

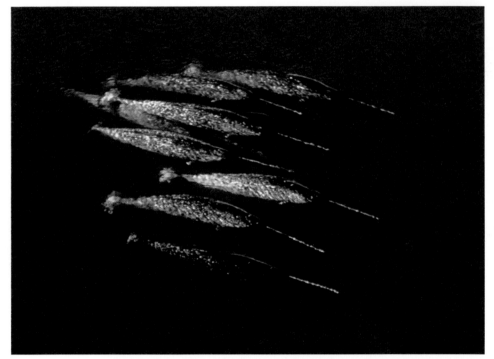

▼Only the male narwhal has a tusk which can be up to 10ft long.

UNICORN OF THE OCEANS

The teeth of the narwhal also seem useless for feeding, although they can be enormous. Females have just two small teeth. Males also have two teeth, but the left one grows into a long pointed tusk. Sometimes both teeth grow. The tusk seems to have no purpose, other than showing that its owner is a male, and occasionally being used in a kind of jousting between males.

HERDS OF HUNDREDS

Throughout the year beluga are in groups, but in mid- to late summer they gather into larger herds and migrate to shallow estuaries. Here the young from the previous year's mating are born. These herds are hundreds or even thousands strong. By virtue of its color, and as most join these big groups, the beluga is relatively easy to count from the air. The world population is about 30,000.

Beluga mothers separate from the herd for a short while to give birth, but soon rejoin a herd. The calf swims so close to the mother that the two are touching. The mother is protective of the calf, which suckles for 2 years.

A MUSEUM PIECE

Commercial hunting of the beluga has greatly reduced its numbers. It is also threatened by shipping activities. The narwhal is often killed for its tusk, which is valued by museums.

▲A newborn beluga is brown. By one year old, like this baby with its mother, it is gray. An adult beluga is white.

▼The forehead "melon" of the beluga is small in babies (1), but has begun to grow at 1 year old (2), and is large by maturity (3) at 5-8 years. Belugas use many sounds and facial expressions to "talk" to one another. At rest the beluga seems to smile (4). A beluga can produce a loud bang by clapping its jaws together (5). A pursed mouth (6) is used in feeding on the sea bottom.

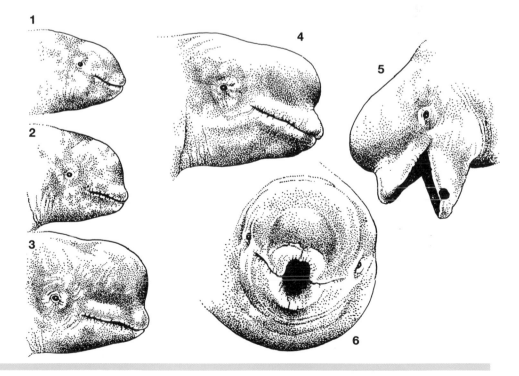

TRUE SEALS

Smooth sleek bodies bask on the shore in the golden sunlight of late afternoon, their mottled fur blending with the rocks. They are Gray seals. Sometimes one of them rolls over and lazily strokes its belly with a flipper. In the swirling sea near by a round gray head with large dark eyes scans the scene before diving out of sight. Below the waves the supple bodies of more Gray seals twist and turn as they chase fish for supper. Exhausted, other Gray seals haul themselves out of the spray to join their sunbathing companions.

Seals are sleek, plump mammals that live in polar and temperate seas and oceans. They are graceful swimmers with short stubby flippers. Their bodies glide easily through the water as they swim. Seals' foreflippers are modified wrists and hands, their hind flippers ankles and feet.

EARLESS SEALS
True seals have no external ear-flaps, and this marks them out from sea lions and fur seals, their close rel-

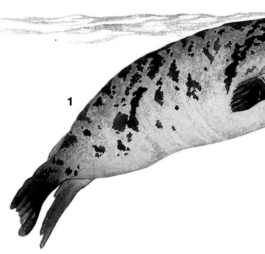

▲ ► Northern and southern true seals
Hooded seal (*Cystophora cristata*) (1). Ringed seal (2). Gray seal (3). Harp seal (4). Bearded seal (*Erignathus barbatus*) (5). Ribbon seal (*Phoca fasciata*) (6). Ross seal (*Ommatophoca rossi*) (7). Weddell seal (8). Crabeater seal (9). Leopard seal (*Hydrurga leptonyx*) (10). Southern elephant seal (11). Hawaiian monk seal (*Monachus schauinslandi*) (12).

▼ Harp seals look out for danger in the Canadian pack ice.

TRUE SEALS Phocidae
(*19 species*)

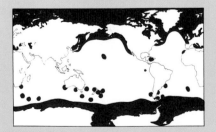

Habitat: offshore rocks and islands, pack-ice, land-fast ice, some large lakes.

Diet: prawns, squid, fish and (larger species of seals) sea-birds and smaller seals.

Breeding: usually 1 young after pregnancy of 10-11 months.

Size: smallest (Ringed seal): head-tail 4ft, weight 100lb; largest (Southern elephant seal): head-tail 16½ft, weight 5,280lb; sexes usually the same size, but males of some species and females of others are much larger.

Color: shades of gray or brown, often with dark spots or patches; young of some species white or tan at birth.

Lifespan: up to 56 years.

Species mentioned in text:
Crabeater seal (*Lobodon carcinophagus*)
Gray seal (*Halichoerus grypus*)
Harbor seal (*Phoca vitulina*)
Harp seal (*P. groenlandica*)
Northern elephant seal (*Mirounga angustirostris*)
Ringed seal (*Phoca hispida*)
Southern elephant seal (*Mirounga leonina*)
Weddell seal (*Leptonychotes weddelli*)

atives. True seals' flippers are weak and they cannot raise their bodies off the ground when on land. A true seal moves on land by using first its chest and then its belly to take its weight.

True seals are more powerful swimmers than sea lions. They use their hind flippers to propel themselves through the water, swinging the back end of their body from side to side.

UNDERWATER HUNTERS

Seals are underwater hunters. They feed on fish, shrimps, squid and sea-birds, which they catch with their pointed teeth. Their large eyes help them see in the dim underwater world.

Sea water is cool, especially at depth, and seals have a thick layer of fat (called blubber) under the skin which prevents them losing heat too quickly. It also helps to give them their smooth shape. The Weddell seal, like other seals, is a good diver, able to go 2,000ft below the surface. It can close its nostrils and hold its breath for more than an hour.

THE SEAL YEAR

Seals breed slowly. The females are not usually ready to breed until they are at least 3 or 4 years old, and they produce only one pup at a time. Seals have to come on to land to give birth, usually in spring or early summer. New-born seals have soft and warm fur, which is not very waterproof.

Female seals (cows) produce some of the richest milk known. In just a few weeks a baby Crabeater seal can grow from 55 to 265lb feeding only on its mother's milk. Suckling may last from 10 days to 12 weeks. Then the cow is ready to mate again.

Male true seals (bulls) often mate with more than one cow. Elephant seals and the Gray seal gather in large numbers on beaches to breed. Each bull defends his own patch of beach and his cows. The bulls roar and slash at each other with their teeth. Weddell seals, Ringed seals and Harbor seals defend underwater territories instead.

After mating, the seals return to the sea to feed. The females are now pregnant, but the babies inside them will not start to develop for 2½ to 3½ months.

▶A pregnant Ringed seal digs out a snow cave above a crack in the pack-ice (inset). Here she will give birth. In this cave, the mother and her pup will be hidden from enemies like the Polar bear and Arctic fox. The snow above will help to keep out the cold.

▶The huge nose of the male Northern elephant seal can be inflated to impress a rival. On his neck are rolls of blubber which may be up to 4in thick.

STILL IN DANGER

In the past, seals have been killed on a large scale for their fur, skins, meat and blubber. Today, international laws limit the numbers that can be killed, and some species are completely protected. But the monk seals are still in danger of becoming extinct, especially in the Mediterranean. This is mainly because the warm coasts where they live are much disturbed by tourists, fishermen and divers.

▼This may look like a family group, but the male Crabeater seal is not the pup's father. He is waiting to mate with its mother.

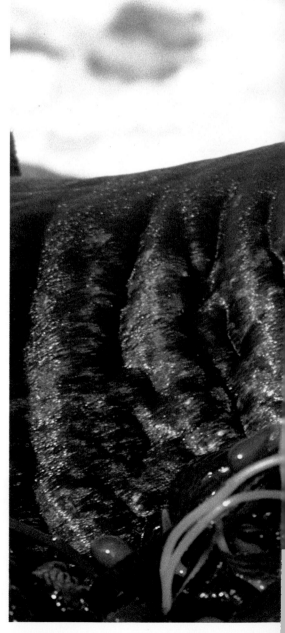

SEA LIONS AND FUR SEALS

Brown shadows weave in and out of the pounding surf, riding the waves and diving back underneath them. California sea lions are playing where the foamy sea meets the land. Beyond the breakers, more sea lions are exploring the sea bed for lobsters and octopus, trailing streams of bubbles behind them. Bull sea lions bark as they argue over ownership of their watery domains.

Sea lions and fur seals are sometimes called eared seals because, unlike the true seals, they have external ear-flaps. The males have thick manes of fur around their necks – the reason for the name sea lions. Fur seals have much thicker fur than sea lions.

SEA LIONS AND FUR SEALS Otariidae (*14 species*)

~~~~ Habitat: offshore rocks and islands.

■ Diet: fish, shrimps, lobsters, octopus and other sea creatures, sometimes sea-birds and young seals.

◯ Breeding: 1 young after pregnancy of 12 months.

Size: smallest (Galapagos fur seal): head-tail 4ft male, weight 60lb; largest (Steller sea lion) head-tail 8½ft male, weight 2,200lb (males larger than females).

Color: shades of brown, gray and tan, males often darker than females, juvenile coat often paler.

Lifespan: up to 25 years.

Species mentioned in text:
Antarctic fur seal (*Arctocephalus gazella*)
Australian sea lion (*Neophoca cinerea*)
California sea lion (*Zalophus californianus*)
Galapagos fur seal (*Arctocephalus galapagoensis*)
Northern fur seal (*Callorhinus ursinus*)
Steller sea lion (*Eumetopias jubatus*)

ACROBATS OF LAND AND SEA

Sea lions are popular performing animals because they are at home on land and in water. They can chase fish underwater or romp around the rocks, tossing fish in the air and catching them in their mouths.

To walk, a sea lion lifts its body off the ground, using its long foreflippers, and swings its hind flippers forwards under its body. When a sea lion wants to go faster over land, it gallops, putting both foreflippers down together, then the hind flippers, then the foreflippers again, and so on. A large fur seal can run faster than a fully grown man.

In the water, sea lions and fur seals are very agile swimmers, but they cannot hold their breath as well as true seals. They rarely dive for more

◀Bull sea lions paddle with their foreflippers as they patrol their water territories.

▼Species of sea lion and fur seal
Males are larger and usually darker than females. Often males have a large mane of thicker fur. Sea lions (1-4) have broader snouts than fur seals (5, 6), which have thicker coats. Male California sea lion (1). Female Steller sea lion (2).

Female South American sea lion (*Otaria avescens*) (3). Male New Zealand sea lion (*Phocarctos hookeri*) (4). Female South American fur seal (*Arctocephalus australis*) (5). Male Northern fur seal (6).

▲ Male California sea lions use ritual threats to argue over territory boundaries on the beach. By using gestures instead of fighting, they have more energy left for mating. Head-shaking and barking as the males approach the boundary (1). Bulls look sideways at each other and make lunges (2). More head-shaking and barking (3). During the lunges, males try to keep their foreflippers away from each other's mouths. The thick skin on their chests softens the blows.

than 5 minutes. Unlike true seals, they use their foreflippers for swimming, flexing their hindquarters for extra power.

BATTLES ON THE BEACHES

In spring and early summer large numbers of sea lions and fur seals gather on their favorite breeding beaches to give birth and mate. The females are heavily pregnant when they arrive and soon give birth. About a week after giving birth, they are ready to mate.

Each male (bull) tries to mate with several females. To compete with other bulls for the females, he tries to defend a section of beach and the females in it. If every dispute led to a fight, the bulls would soon become too exhausted to mate. Instead, they make threatening displays and gestures, from which they can usually judge which animal is the stronger.

For a bull to defend his territory throughout the breeding season, he must stay on his patch. So most bulls do not feed at all during this time. Sometimes they may fast for 70 days. They can do this because they live off their fat (blubber). The biggest bulls have the most fat, so they are usually the most successful in holding a territory. The weather is usually warm at this time, and the seals get very hot on the beach. So the most prized territories are those nearest the water.

◄ This Australian sea lion pup will soon shed his two-color coat for a dark-brown adult one.

WELL-FED PUPS

Mother sea lions and fur seals stay with their pups for the first week of their lives, suckling them frequently. The pups need to be protected from the bulls, which can easily trample them during a fight.

As each pup grows bigger, its mother spends longer at sea feeding, returning from time to time to suckle her young. She finds her pup by calling to it and listening for its answering call. Many pups do not leave their mother until the next pup arrives 1 to 3 years later.

The Northern fur seal migrates hundreds of miles to different feeding areas in summer and winter. Its pups stop suckling when the migration starts. Female pups reach maturity at about 4 years old and male pups at 5 to 8 years of age.

PROTECTED POPULATIONS

Instead of having mainly long coarse hairs with just a few shorter ones, like the sea lions, the fur seals have a dense layer of woolly underfur. Glands among the hairs keep the animals coated with waterproof oil.

The thick fur causes problems in summer, when fur seals suffer badly from the heat. The only part of their bodies that can lose heat is the flippers. The animals often wave their flippers in the air to cool themselves.

But the fur seals' coats have caused them bigger problems than this. They have been much sought by hunters. By the end of the nineteenth century so many fur seals had been killed that some species were almost extinct. Laws were later passed to prevent the slaughter. Exploitation continues, but under strict international controls.

During the twentieth century fur seals have made a strong recovery. The Antarctic fur seal, whose population was reduced to probably fewer than 50 animals, now numbers around a million, and the population is increasing. Scientists think that this is partly because so many whales have been killed. The whales used to compete with the seals for one of their favorite foods, the tiny shrimp-like krill.

The Australian sea lion is the best example of the improved relationship between seals and people. In places, it is so accustomed to human visitors that tourists can mingle with the seals on the beach.

▼An Antarctic fur seal with a plastic packing band cutting into its neck. Harmful waste kills many seals.

WALRUS

With a grunt and a splash the large bristly snout of a walrus comes out of the water. It is followed by 2,200lb of brownish-pink flesh. Using his long pointed tusks as levers, the walrus hauls himself out of the sea on to an ice-floe. He shuffles towards a group of dozing walruses and flops down on the ice to snooze, using the belly of another walrus as a pillow.

The walrus is rather like a giant pink sea lion. It has flippers instead of legs, and a fat spindle-shaped body. It is a powerful swimmer and can stay at sea for up to 2 days at a time. When on land, the walrus props up its body on its foreflippers, tucking its hind flippers underneath. It walks awkwardly, shuffling along on its flippers.

The females (cows) and young males have short velvety coats. Adult males (bulls) have little hair and look naked. Their skin is up to 2in thick, wrinkled like an old leather bag.

The walrus has such a short thick neck that its head seems to be joined directly to its shoulders. Its squarish snout is covered in stiff bristles. One pair of teeth in the upper jaws of both sexes form tusks up to 22in long.

WHY HAVE TUSKS?

The walrus does not use its tusks for feeding. It feeds on the sea bed, using the bristles on its snout to feel for clams, mussels and other sea creatures. It digs in the mud with its snout, which is covered in extra tough skin. Sometimes it shoots a jet of water from its mouth to blast prey animals out of their burrows.

The walrus uses its tusks like ice-axes to haul itself on to ice-floes, or to smash breathing-holes in the ice.

A walrus also uses its tusks to establish its place in the group. Walruses usually live in very large groups, sometimes of several thousand animals, and there are often arguments. The male walruses with the largest tusks get the best places on the ice or beach and the best chance of mating with the females. Walruses display to each other, showing off their tusks. If a display does not settle a dispute, the walruses may stab at each other with their tusks.

For thousands of years, Eskimos have hunted walruses for their skins and meat. Where stocks allow, this continues today.

▶Sunbathing walruses look pink as their blood flows to the surface to absorb the Sun's warmth. The walrus just leaving the water is much paler, its blood flowing deeper to avoid losing heat to the cold water.

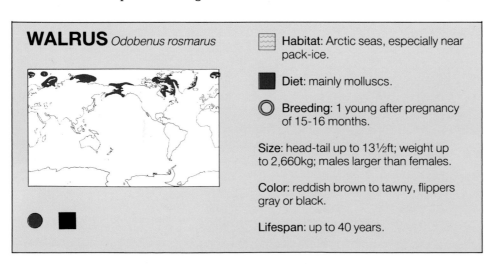

WALRUS *Odobenus rosmarus*

Habitat: Arctic seas, especially near pack-ice.

Diet: mainly molluscs.

Breeding: 1 young after pregnancy of 15-16 months.

Size: head-tail up to 13½ft; weight up to 2,660kg; males larger than females.

Color: reddish brown to tawny, flippers gray or black.

Lifespan: up to 40 years.

GLOSSARY

Adaptation Features of an animal's body or life-style that suit it to its environment.

Aggression Behaviour in which one animal attacks or threatens another.

Aquatic Living for much, if not all, of the time in the water.

Baleen whales Whales that feed on small marine life by filtering them through horny plates of baleen. *Compare* Toothed whales.

Bear family The family Ursidae, which includes the largest land carnivores, the Grizzly and Polar bears. It also includes the Black bear, the Sun bear and Sloth bear.

Big cats Members of the genus *Panthera*, including the lion, tiger, leopard, jaguar and cheetah.

Blubber The thick layer of fat beneath a whale's skin.

Camouflage Colour and patterns on an animal's coat that allow it to blend in with its surroundings.

Canids Animals belonging to the dog family, Canidae.

Carnivore In general, an animal that eats meat. Specifically, a member of the order Carnivora, which includes the cat, dog, bear, raccoon, weasel, mongoose and hyena families.

Carrion The meat from a dead animal. It forms a major part of the diet of some carnivores, such as hyenas and jackals.

Cat family The family of carnivores Felidae, which includes the big cats such as the lion, tiger, leopard, jaguar and cheetah; and the small cats such as the lynx, bobcat, wildcat, ocelot and the domestic cat.

Cetaceans Members of the order Cetacea, which include whales (Baleen and Toothed), dolphins and porpoises.

Class The division of animal classification above Order.

Competition The contest between two or more species over such things as space and food.

Conservation Looking after living things, their habitat and the environment in general.

Dasyurids Members of the family Dasyuridae, which are carnivorous marsupials.

Den A shelter in which an animal or group of animals sleeps, hides, or gives birth to young.

Diet The food an animal eats.

Digit A finger or a toe.

Diurnal Active during the day.

Dog family The family Canidae, whose 35 species include the wolves, coyote, dingo, foxes, jackals, wild dogs and dhole.

Dominant Of higher rank; a dominant male of a group is "the boss".

Echo-location The method dolphins, porpoises and other toothed whales use to find food. They send out high-pitched sound signals and listen for echoes which the prey will reflect back.

Ecology The study of living things in relation to their environment.

Endangered species One whose numbers have dropped so low that it is in danger of becoming extinct.

Environment The surroundings, of a particular species, or the world about us in general.

Extinction The complete loss of a species, either locally or on the Earth.

Family The division of animal classification below Order and above Genus.

Felids Members of the cat family, Felidae, including the big cats and the small cats.

Feral Leading a wild existence; usually refers to a formerly domesticated animal, such as a cat, that has escaped from captivity.

Genus The division of animal classification below Family and above Species.

Gestation The period of pregnancy of an animal.

Helper An animal that assists in raising offspring that are not its own. This practice is notable in jackals, for example.

Home range The area in which an animal usually lives and feeds.

Hyenids Members of the hyena family, Hyaenidae.

Mammals A class of animals whose females have mammary glands, which produce milk on which they feed their young.

Marine Living in the sea.

Marsupials An order of mammals, whose females give birth to very under-developed young and then raise them (usually) in a pouch.

Migration The long-distance movement of animals, usually seasonal, for the purposes of feeding or breeding.

Moongoose family The family Viverridae, which includes mongooses, civets and linsangs.

Mustelids Members of the weasel family, Mustelidae.

Nocturnal Active during the night.

Nomadic Leading a wandering life, with no distinct territory. Wolf packs on the northern tundra are nomadic, following the herds of caribou on which they prey.

Omnivore An animal that has a varied diet, including both plants and animals.

Order The division of animal classification below Class and above Family.

Pack A group of wolves or other dogs. A wolf pack can have up to 20 animals.

Pinnipeds Members of the order Pinnipedia, which includes seals, sea lions and the walrus. The word pinniped means wing-footed. It refers to the limbs of these animals, which are flippers.

Pod Term for a group of whales, occasionally of other animals.

Population A separate group of animals of the same species.

Predator An animal that hunts live prey.

Pregnancy Period during which the young grows inside the body of a mammal.

Prey An animal hunted for food by a predator.

Pride The social unit of lions. It consists of a group of up to 12 related adult females and their offspring and up to six adult males.

Procyonids Members of the raccoon family, Procyonidae.

Rabies A deadly virus disease that affects people and virtually all warm-blooded animals. It can be spread by most carnivores through the bite. Dog bites are the usual cause of human infection. Another term for rabies is hydrophobia, meaning fear of water. People infected with rabies cannot drink water.

Raccoon family The family Procyonidae, which includes raccoons, coatis and pandas.

Race The division of animal classification below sub-species; it refers to animals that are very similar but have slightly different characteristics, eg Highland and Lowland gorillas.

Retractile Refers to claws, which are able to be withdrawn into protective sheaths, as by members of the cat family.

Savannah The tropical grassland of Africa, Central and South America and Australia.

Scavenger An animal that feeds on the remains of carcasses that others have abandoned. Hyenas are master scavengers, crushing and eating bones, from which they can extract every bit of organic matter.

Scent marking Marking territory by smearing objects with scent from glands in the body or by means of urine.

Small cats Members of the cat family belonging to the genus *Felis*. They

include the lynx, bobcat, puma, wild cat and ocelot.

Solitary Living alone for most of the time.

Species The division of animal classification below Genus; a group of animals of the same structure which can breed with one another.

Steppe The temperate grassy plains of Eurasia. Called "prairie" in North America.

Sub-species The division of animal classification below Species and above Race; typically the sub-species are separated geographically.

Terrestrial Spending most of the time on the ground.

Territory The area in which an animal or group of animals lives and defends against intruders.

Toothed whales Whales belonging to the suborder Odontoceti, which accounts for nearly 90 per cent of all whales. They comprise the dolphins, including the Killer whale, porpoises, white whales, sperm whales and beaked whales. *Compare* Baleen whales.

Tundra The barren treeless land in the far north of Europe, Asia and North America. The vegetation includes low shrub, moss and lichen.

Ursids Members of the bear family, Ursidae.

Viverrids Members of the mongoose family, Viverridae.

Weasel family Animals belonging to the family Mustelidae, which include weasels, polecats, ferrets, martens, skunks, otters, badgers, minks and zorilla.

INDEX

Common names

Single page numbers indicate a major section of the main text. Double, hyphenated, numbers refer to major articles. **Bold numbers** refer to illustrations.

aardwolf **65**
American black bear 38-39
antechinus
   Brown 66, 68
   Dusky **66**
   Fat-tailed **69**
   Little red **69**
   Red-eared *see* antechinus,
     Fat-tailed

badger
   American 58, **59**
   Eurasian **58**, 59
   Oriental ferret 58, **59**
badgers 58-59
   see also badger, ratel,
     teledu
bear
   American black 38-39
   Asian black **41**
   Grizzly (Brown) 36-37, 39
   Polar 34-35
   Sloth **40**, **41**
   Spectacled 40, **41**
   Sun 40, 41
bears
   small 40-41
   see also bear
beluga 78, **79**
binturong *see* cat, Bear
bobcat **20**

cat
   Asiatic golden **19**
   Bear **60**
   Black-footed **19**
   Domestic 18, 20
   Fishing **19**
   Geoffroy's 20
   Jungle **19**
   Leopard 18, **19**
   Margay **19**, 20
   New Guinea marsupial **69**
   Sand **19**
   Tiger **19**
cats
   Feral 21
   see also cat
cheetah 16-17
civet
   African 60
   African palm 60
   Banded palm **60**
   Celebes palm *see* civet,

Giant palm
Common palm **60**
Giant palm **60**
Jerdon's palm 60
Large spotted 60
Oriental **60**
Otter 60
civets 60-61
   see also cat, Bear; civet;
     linsang, African
coati
   Ring-tailed 42, **43**
   White-nosed 42, **43**
coatis 42-43
   see also coati
cougar *see* puma
coyote 24-25

devil
   Tasmanian 66, **67**, 68
dhole 33
dibbler 68
dingo **33**
dog
   Bush 33
   Domestic 23, 33
   Raccoon 33
   Small-eared 28, **30**, 31
   Wild *see* wild dog
dolphin
   Atlantic humpbacked **70**
   Atlantic white-sided **70**
   Bottle-nosed **70**, 72
   Commerson's **70**
   Common **70**
   Dusky **70**, 72
   Heaviside's 70
   Northern right whale **70**
   Risso's **70**
   Rough-toothed **70**
   Spotted **70**
dolphins 70-73
   see also dolphin; whale,
     Falsekiller; whale, Killer;
     whale, Melon-headed
dunnart
   Fat-tailed 68, **69**

ermine *see* stoat

ferret
   Black-footed **47**
fisher 50
fox
   Arctic **30**, 31
   Argentine gray **30**, 31
   Azara's **30**
   Bat-eared 30, **31**
   Blanford's **28**
   Cape **28**
   Colpeo **30**, 31
   Corsac **28**

Crab-eating **30**
Fennec **28**, 30
Gray **28**, 29, 31
Indian **28**, 29
Kit **28**, **31**
Red 28, **29**, 30, 31
Rüppell's **28**
Swift *see* fox, Kit
Tree *see* fox, Gray
foxes 28-31
   see also Fox
fur seal
   Antarctic **87**
   Galapagos 84
   Northern **85**, 87
   South American **85**
fur seals 84-87
   see also fur seal

genet
   Common 60
genets 60
   see also genet
grison 46
   Little **47**
Grizzly (Brown) bear 36-37

hyena
   Brown **64**, **65**
   Spotted **64**, **65**
   Striped **64**, **65**
hyenas 64-65
   see also aardwolf, hyena

jackal
   Golden 26, **27**
   Sidestriped 26, **27**
   Silverbacked **26**, **27**
   Simien 26, **27**
jackals 26-27
   see also jackal
jaguar 12-15
jaguarundi **19**

kultarr **69**

leopard 12, **13**, 14
   Clouded **12**, 15
   Snow **12**, 14, **15**
leopards 12-15
   see also leopard
linsang
   African **60**
linsangs 60
   see also linsang
lion 6-9
   Mountain *see* puma
lynx 19, 20

marsupial carnivores 66-69
marten
   American 50

Beech *see* marten, Stone
Pine **50**
Stone **50**
martens 50-51
   see also fisher, marten,
     sable
meercat
   Gray *see* suricate
mongoose
   Banded 62, **63**
   Dwarf 62, **63**
   Egyptian **63**
   Marsh **63**
   Narrow-striped **63**
   Ring-tailed **63**
   Selous' **63**
   Small Indian 63
   White-tailed, 62, **63**
mongooses 62-63
   see also mongoose,
     suricate
mouse
   Long-tailed marsupial **69**
   Marsupial **69**
   Three-striped marsupial
     69
mulgara 66, 67, 68

narwhal 78, 79
ningaui
   Pilbara 66, **69**

ocelot **19**, 20
otter
   Cape clawless **57**
   European river 56
   Giant 56, **57**
   Indian smooth-coated **57**
   North American river 56, **57**
   Oriental short-clawed **56**,
     57
   Sea 56
   Spot-necked **57**
otters 56-57 *see also* otter
ounce *see* leopard, Snow

panther
   black *see* leopard
phascogale
   Red-tailed **69**
phascogales 68
   see also phascogale
planigale
   Common 66, **69**
Polar bear 34-35
polecat
   European **46**, **47**, **48**
   Marbled, **47**
polecats 46-49
   see also ferret, grison,
     polecat, stoat, weasel,
     zorilla

Scientific names

The first name of each double-barrel Latin name refers to the *Genus*, the second to the *species*. Single names not in *italic* refer to a family or sub-family and are cross-referenced to the Common name index.

FURTHER READING

Alexander, R. McNeill (ed)(1986), *The Encyclopedia of Animal Biology*, Facts on File, New York

Berry, R.J. and Hallam, A. (eds)(1986), *The Encyclopedia of Animal Evolution*, Facts on File, New York

Bertram, B.C. (1978), *Pride of Lions*, Charles Scribner, New York

Corbet, G.B. and Hill, J.E. (1980), *A World List of Mammalian Species*, British Museum and Cornell University Press, London and Ithaca, NY

Fox, M.W. (ed)(1975), *The Wild Canids: their Systematics, Behavioral Ecology, Evolution*, Van Nostrand Reinhold, London and New York

Griffiths, M.E. (1978), *The Biology of Monotremes*, Academic Press, New York

Grzimek, B. (ed)(1972), *Grzimek's Animal Life Encyclopedia*, vols 10, 11, 12, Van Nostrand Reinhold, New York

Hall, E.R. and Kelson, K.R. (1959), *The Mammals of North America*, Ronald Press, New York

Harrison Matthews, L. (1969), *The Life of Mammals*, vols 1 and 2, Weidenfeld and Nicolson, London

Hunsaker II, D. (ed)(1977), *The Biology of Marsupials*, Academic Press, New York

King, J.E. (1983), *Seals of the World*, Oxford University Press, England

Kingdon, J. (1971-82), *East African Mammals*, vols I-III, Academic Press, New York

Lawick, H. van and J. van Lawick-Goodall (1970), *The Innocent Killers*, Collins, London

Macdonald, D. (ed)(1984) *The Encyclopedia of Mammals*, Facts on File, New York

Moore, P.D. (ed)(1986), *The Encyclopedia of Animal Ecology*, Facts on File, New York

Nowak, R.M. and Paradiso, J.L. (eds)(1983) *Walker's Mammals of the World* (4th edn) 2 vols, Johns Hopkins University Press, Baltimore and London

Pelton, M.R., Lentfer, J.W. and Stokes, G.E. (eds)(1976), *Bears: their Biology and Management*, IVCN Publ, New Series no.40, Morges, Switzerland

Slater, P.J.B. (ed)(1986), *The Encyclopedia of Animal Behavior*, Facts on File, New York

Strahan, R. (ed), *The Complete Book of Australian Mammals*, Angus and Robertson, Sydney

Winn, H.E. and Olla, B.L. (1979), *The Behavior of Marine Mammals*, vol 3, Cetaceans, Plenum, New York

Young, J.Z. (1975), *The Life of Mammals: their Anatomy and Physiology*, Oxford University Press, Oxford

ACKNOWLEDGMENTS

Picture credits

Key: *t* top *b* bottom *c* centre *l* left *r* right
Abbreviations: A Ardea. AH Andrew Henley. AN Natrue, Agence Photographique. BC Bruce Coleman Ltd. FL Frank Lane Agency. FS Fiona Sunquist. GF George Frame. J Jacana. OSF Oxford Scientific Films.

6 A. 7 GF. 8 BC. 10*t* BC, 10*b* FS. 13*t* GF, 13*b* GF. 14 A. 15*t* AN, 15*b* BC/R. Williams. 17*t* GF, 17*b* GF. 19 Natural Science Photos. 20/21 BC/Leonard Lee Rue III. 23 A. 24 OSF. 25*t* OSF, 25*b* BC. 26 R. Caputo. 28 OSF. 29 R.O. Peterson. 31*t* FS, 31*b* A. 32 L. Malcolm. 33*t* BC, 33*b* Auscape International/J.P. Ferro. 34 J.W. Lentfer. 35*t* FL/C. Carvalho, 35*b* BC. 36 Leonard Lee Rue III. 37 OSF. 38 OSF. 39*t* A, 39*b* BC. 40 J. Mackinnon. 42 BC. 43 BC. 44*l* BC, 44*r* BC. 45*t* OSF, 44*b* FL. 46 BC. 48/49 Biofotos/G.Kinns. 51 Zefa. 52 S. Carlsson. 53 AN. 55 BC. 56 J. 57 BC. 58 OSF. 59 Leonard Lee Rue III. 60 J. Mackinnon. 63 D. Macdonald. 64 J. 65 GF. 66 AH. 67*t* AH, 67*bl* A, 67*r* WNF Switzerland. 72 AN. 73 M. Wursig. 75 D. Gaskin. 76, 77 Sea Mammal Research Unit, Cambridge. 78 F. Bruemmer. 80 N.R. Lightfoot. 82 R.M. Laws. 82/83 A. 84 J. 86 AH. 87 W.N. Bonner. 88/89 FL.

Artwork credits

Key: *t* top *b* bottom *c* centre *l* left *r* right.
Abbreviations: OI Oxford Illustrators. PB Priscilla Barrett.

10/11 PB. 12*l* PB *r* Michael Long. 16/17*b* Denys Ovenden, 17*t* Mick Saunders. 18/19 PB. 22 PB. 23 PB. 25 PB. 27 PB. 28/29 PB. 30 PB. 31 PB. 32 PB. 38 Denys Ovenden. 41 PB. 43 PB. 45 Denys Ovenden. 46 PB. 47 PB. 50 Rob van Assen. 53 PB. 54*l* VAP, 54*r* PB. 55 PB. 57 PB. 59 PB. 61 PB. 62/63 PB. 63 PB. 64 PB. 65 PB. 68/69 Dick Twinney. 70/71 Malcolm McGregor. 74 Malcolm McGregor. 76/77 Malcolm McGregor. 77*r* Stephen Cocking. 78 Malcolm McGregor. 79*t* Malcolm McGregor, 79*b* PB. 80/81 PB. 82/83 Simon Driver. 85 PB. 86 PB.